SCHAUM'S *Easy* O

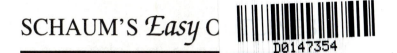

DIFFERENTIAL
EQUATIONS

Other Books in Schaum's Easy Outlines Series Include:

SCHAUM'S *Easy* OUTLINES

DIFFERENTIAL EQUATIONS

BASED ON SCHAUM'S
Outline of Theory and Problems of
Differential Equations, Second Edition
BY RICHARD BRONSON, Ph.D.

ABRIDGEMENT EDITOR
ERIN J. BREDENSTEINER, Ph.D.

SCHAUM'S OUTLINE SERIES

McGRAW-HILL

New York   Chicago   San Francisco   Lisbon   London   Madrid
Mexico City   Milan   New Delhi   San Juan
Seoul   Singapore   Sydney   Toronto

The McGraw·Hill Companies

RICHARD BRONSON, Ph.D is Professor of Mathematics at Fairleigh Dickinson University. He received his Ph.D. in applied mathematics from Stevens Institute of Technology in 1968. He has served as an associate editor of the journal *Simulation*, as a contributing editor to *SIAM News*, and as a consultant to Bell Laboratories. He has published over 30 technical articles and books, including *Schaum's Outline of Matrix Operations* and *Schaum's Outline of Operations Research*.

ERIN J. BREDENSTEINER, Ph.D teaches mathematics at the University of Evansville in Indiana. She received her B.S., M.S., and Ph.D. degrees in mathematics from Rensselaer Polytechnic Institute, and she is the co-author of several journal articles.

4 5 6 7 8 9 0 DOC/DOC 0 9 8 7 6

ISBN 0-07-140967-X

Contents

Chapter 1
BASIC CONCEPTS AND CLASSIFYING DIFFERENTIAL EQUATIONS

IN THIS CHAPTER:

- ✔ *Differential Equations*
- ✔ *Notation*
- ✔ *Solutions*
- ✔ *Initial-Value and Boundary-Value Problems*
- ✔ *Standard and Differential Forms*
- ✔ *Linear Equations*
- ✔ *Bernoulli Equations*
- ✔ *Homogeneous Equations*
- ✔ *Separable Equations*
- ✔ *Exact Equations*

Differential Equations

A *differential equation* is an equation involving an unknown function and its derivatives.

Example 1.1: The following are differential equations involving the unknown function y.

$$\frac{dy}{dx} = 5x + 3 \tag{1.1}$$

$$e^y \frac{d^2y}{dx^2} + 2\left(\frac{dy}{dx}\right)^2 = 1 \tag{1.2}$$

$$4\frac{d^3y}{dx^3} + (\sin x)\frac{d^2y}{dx^2} + 5xy = 0 \tag{1.3}$$

$$\left(\frac{d^2y}{dx^2}\right)^3 + 3y\left(\frac{dy}{dx}\right)^7 + y^3\left(\frac{dy}{dx}\right)^2 = 5x \tag{1.4}$$

$$\frac{\partial^2 y}{\partial t^2} - 4\frac{\partial^2 y}{\partial x^2} = 0 \tag{1.5}$$

A differential equation is an *ordinary differential equation* if the unknown function depends on only one independent variable. If the unknown function depends on two or more independent variables, the differential equation is a *partial differential equation. In this book we will be concerned solely with ordinary differential equations.*

Example 1.2: Equations 1.1 through 1.4 are examples of ordinary differential equations, since the unknown function y depends solely on the variable x. Equation 1.5 is a partial differential equation, since y depends on both the independent variables t and x.

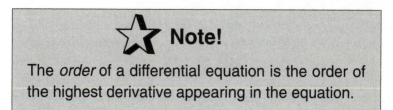

Note!

The *order* of a differential equation is the order of the highest derivative appearing in the equation.

Example 1.3: Equation 1.1 is a first-order differential equation; 1.2, 1.4, and 1.5 are second-order differential equations. (Note in 1.4 that the order of the highest derivative appearing in the equation is two.) Equation 1.3 is a third-order differential equation.

Notation

The expressions $y', y'', y''', y^{(4)}, ..., y^{(n)}$ are often used to represent, respectively, the first, second, third, fourth, ..., nth derivatives of y with respect to the independent variable under consideration. Thus, y'' represents $d^2 y / dx^2$ if the independent variable is x, but represents $d^2 y / dp^2$ if the independent variable is p. Observe that parenthesis are used in $y^{(n)}$ to distinguish it from the nth power, y^n. If the independent variable is time, usually denoted by t, primes are often replaced by dots. Thus, $\dot{y}, \ddot{y},$ and $\dddot{y}$ represent, $dy / dt, d^2 y / dt^2$, and $d^3 y / dt^3$, respectively.

Solutions

A *solution* of a differential equation in the unknown function y and the independent variable x on the interval $\mathcal{I}$ is a function $y(x)$ that satisfies the differential equation identically for all x in $\mathcal{I}$.

Example 1.4: Is $y(x) = c_1 \sin 2x + c_2 \cos 2x$, where c_1 and c_2 are arbitrary constants, a solution of $y'' + 4y = 0$?

Differentiating y, we find $y' = 2c_1 \cos 2x - 2c_2 \sin 2x$ and $y'' = -4c_1 \sin 2x - 4c_2 \cos 2x$. Hence,

$$\begin{aligned} y'' + 4y &= (-4c_1 \sin 2x - 4c_2 \cos 2x) + 4(c_1 \sin 2x + c_2 \cos 2x) \\ &= (-4c_1 + 4c_1)\sin 2x + (-4c_2 + 4c_2)\cos 2x \\ &= 0 \end{aligned}$$

Thus, $y = c_1 \sin 2x + c_2 \cos 2x$ satisfies the differential equation for all values of x and is a solution on the interval $(-\infty, \infty)$.

Example 1.5: Determine whether $y = x^2 - 1$ is a solution of $(y')^4 + y^2 = -1$.

Note that the left side of the differential equation must be nonnegative for every real function $y(x)$ and any x, since it is the sum of terms raised to the second and fourth powers, while the right side of the equation is negative. Since no function $y(x)$ will satisfy this equation, the given differential equation has no solutions.

We see that some differential equations have infinitely many solutions (Example 1.4), whereas other differential equations have no solutions (Example 1.5). It is also possible that a differential equation has exactly one solution. Consider $(y')^4 + y^2 = 0$, which for reasons identical to those given in Example 1.5 has only one solution $y \equiv 0$.

You Need to Know ✔

A *particular solution* of a differential equation is any one solution. The *general solution* of a differential equation is the set of all solutions.

Example 1.6: The general solution to the differential equation in Example 1.4 can be shown to be (see Chapters Four and Five) $y = c_1 \sin 2x + c_2 \cos 2x$. That is, every particular solution of the differential equation has this general form. A few particular solutions are: (a) $y = 5\sin 2x - 3\cos 2x$ (choose $c_1 = 5$ and $c_2 = -3$), (b) $y = \sin 2x$ (choose $c_1 = 1$ and $c_2 = 0$), and (c) $y \equiv 0$ (choose $c_1 = c_2 = 0$).

The general solution of a differential equation cannot always be expressed by a single formula. As an example consider the differential equation $y' + y^2 = 0$, which has two particular solutions $y = 1/x$ and $y \equiv 0$.

Initial-Value and Boundary-Value Problems

A differential equation along with subsidiary conditions on the unknown function and its derivatives, all given at the same value of the independent variable, constitutes an *initial-value problem*. The subsidiary conditions are *initial conditions*. If the subsidiary conditions are given at more than one value of the independent variable, the problem is a *boundary-value problem* and the conditions are *boundary conditions*.

Example 1.7: The problem $y'' + 2y' = e^x; y(\pi) = 1, y'(\pi) = 2$ is an initial value problem, because the two subsidiary conditions are both given at $x = \pi$. The problem $y'' + 2y' = e^x; y(0) = 1, y(1) = 1$ is a boundary-value problem, because the two subsidiary conditions are given at $x = 0$ and $x = 1$.

A solution to an initial-value or boundary-value problem is a function $y(x)$ that both solves the differential equation and satisfies all given subsidiary conditions.

Standard and Differential Forms

Standard form for a first-order differential equation in the unknown function $y(x)$ is

$$y' = f(x, y) \tag{1.6}$$

where the derivative y' appears only on the left side of 1.6. Many, but not all, first-order differential equations can be written in standard form by algebraically solving for y' and then setting $f(x,y)$ equal to the right side of the resulting equation.

The right side of 1.6 can always be written as a quotient of two other functions $M(x,y)$ and $-N(x,y)$. Then 1.6 becomes $dy / dx = M(x,y) / -N(x,y)$, which is equivalent to the *differential form*

$$M(x,y)dx + N(x,y)dy = 0 \tag{1.7}$$

Linear Equations

Consider a differential equation in standard form 1.6. If $f(x,y)$ can be written as $f(x,y) = -p(x)y + q(x)$ (that is, as a function of x times y, plus another function of x), the differential equation is linear. First-order linear differential equations can always be expressed as

$$y' + p(x)y = q(x) \tag{1.8}$$

Linear equations are solved in Chapter Two.

Bernoulli Equations

A Bernoulli differential equation is an equation of the form

$$y' + p(x)y = q(x)y^n \tag{1.9}$$

where n denotes a real number. When $n = 1$ or $n = 0$, a Bernoulli equation reduces to a linear equation. Bernoulli equations are solved in Chapter Two.

Homogeneous Equations

A differential equation in standard form (1.6) is *homogeneous* if

$$f(tx,ty) = f(x,y) \tag{1.10}$$

for every real number t. Homogeneous equations are solved in Chapter Two.

⭐ **Note!**

In the general framework of differential equations, the word "homogeneous" has an entirely different meaning (see Chapter Four). Only in the context of first-order differential equations does "homogeneous" have the meaning defined above.

Separable Equations

Consider a differential equation in differential form (1.7). If $M(x,y) = A(x)$ (a function only of x) and $N(x,y) = B(y)$ (a function only of y), the differential equation is *separable*, or has its *variables separated*. Separable equations are solved in Chapter Two.

Exact Equations

A differential equation in differential form (1.7) is exact if

$$\frac{\partial M(x,y)}{\partial y} = \frac{\partial N(x,y)}{\partial x} \qquad (1.11)$$

Exact equations are solved in Chapter Two (where a more precise definition of exactness is given).

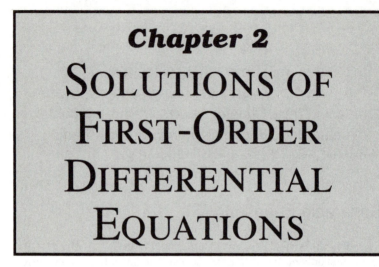

Chapter 2

SOLUTIONS OF FIRST-ORDER DIFFERENTIAL EQUATIONS

IN THIS CHAPTER:

- ✔ *Separable Equations*
- ✔ *Homogeneous Equations*
- ✔ *Exact Equations*
- ✔ *Linear Equations*
- ✔ *Bernoulli Equations*
- ✔ *Solved Problems*

Separable Equations

General Solution

The solution to the first-order separable differential equation (see Chapter One).

$$A(x)dx + B(y)dy = 0 \qquad (2.1)$$

is

$$\int A(x)dx + \int B(y)dy = c \qquad (2.2)$$

where c represents an arbitrary constant.
(See Problem 2.1)
The integrals obtained in Equation 2.2 may be, for all practical purposes, impossible to evaluate. In such case, numerical techniques (see Chapter 14) are used to obtain an approximate solution. Even if the indicated integrations in 2.2 can be performed, it may not be algebraically possible to solve for y explicitly in terms of x. In that case, the solution is left in implicit form.

Solutions to the Initial-Value Problem

The solution to the initial-value problem

$$A(x)dx + B(y)dy = 0; \quad y(x_0) = y_0 \qquad (2.3)$$

can be obtained, as usual, by first using Equation 2.2 to solve the differential equation and then applying the initial condition directly to evaluate c.

Alternatively, the solution to Equation 2.3 can be obtained from

$$\int_{x_0}^{x} A(s)ds + \int_{y_0}^{y} B(t)dt = 0 \qquad (2.4)$$

where s and t are variables of integration.

Homogeneous Equations

The homogeneous differential equation

$$\frac{dy}{dx} = f(x,y) \qquad (2.5)$$

having the property $f(tx, ty) = f(x, y)$ (see Chapter One) can be transformed into a separable equation by making the substitution

$$y = xv \qquad (2.6)$$

along with its corresponding derivative

$$\frac{dy}{dx} = v + x\frac{dv}{dx} \qquad (2.7)$$

The resulting equation in the variables v and x is solved as a separable differential equation; the required solution to Equation 2.5 is obtained by back substitution.

Alternatively, the solution to 2.5 can be obtained by rewriting the differential equation as

$$\frac{dx}{dy} = \frac{1}{f(x,y)} \qquad (2.8)$$

and then substituting

$$x = yu \qquad (2.9)$$

and the corresponding derivative

$$\frac{dx}{dy} = u + y\frac{du}{dy} \qquad (2.10)$$

into Equation 2.8. After simplifying, the resulting differential equation will be one with variables (this time, u and y) separable.

Ordinarily, it is immaterial which method of solution is used. Occasionally, however, one of the substitutions 2.6 or 2.9 is definitely superior to the other one. In such cases, the better substitution is usually apparent from the form of the differential equation itself.

(See Problem 2.2)

Exact Equations

Defining Properties

A differential equation

$$M(x, y)dx + N(x, y)dy = 0 \qquad (2.11)$$

is *exact* if there exists a function $g(x, y)$ such that

$$dg(x, y) = M(x, y)dx + N(x, y)dy \qquad (2.12)$$

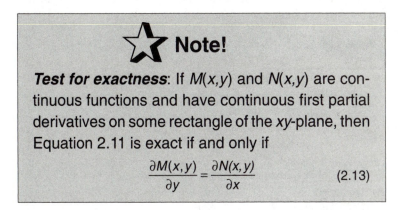

⭐ Note!

Test for exactness: If $M(x,y)$ and $N(x,y)$ are continuous functions and have continuous first partial derivatives on some rectangle of the xy-plane, then Equation 2.11 is exact if and only if

$$\frac{\partial M(x,y)}{\partial y} = \frac{\partial N(x,y)}{\partial x} \qquad (2.13)$$

Method of Solution

To solve Equation 2.11, assuming that it is exact, first solve the equations

$$\frac{\partial g(x,y)}{\partial x} = M(x,y) \qquad (2.14)$$

$$\frac{\partial g(x,y)}{\partial y} = N(x,y) \qquad (2.15)$$

for $g(x, y)$. The solution to 2.11 is then given implicitly by

$$g(x, y) = c \qquad (2.16)$$

where c represents an arbitrary constant.

Equation 2.16 is immediate from Equations 2.11 and 2.12. If 2.12 is substituted into 2.11, we obtain $dg(x, y(x)) = 0$. Integrating this equation (note that we can write 0 as $0\,dx$), we have $\int dg(x, y(x)) = \int 0\,dx$, which, in turn, implies 2.16.

Integrating Factors

In general, Equation 2.11 is not exact. Occasionally, it is possible to transform 2.11 into an exact differential equation by a judicious multiplication. A function $I(x, y)$ is an *integrating factor* for 2.11 if the equation

$$I(x, y)[M(x, y)dx + N(x, y)dy] = 0 \tag{2.17}$$

is exact. A solution to 2.11 is obtained by solving the exact differential equation defined by 2.17. Some of the more common integrating factors are displayed in Table 2.1 and the conditions that follow:

If $\dfrac{1}{N}\left(\dfrac{\partial M}{\partial y} - \dfrac{\partial N}{\partial x} \right) \equiv g(x)$, a function of x alone, then

$$I(x, y) = e^{\int g(x)dx} \tag{2.18}$$

If $\dfrac{1}{M}\left(\dfrac{\partial M}{\partial y} - \dfrac{\partial N}{\partial x} \right) \equiv h(y)$, a function of y alone, then

$$I(x, y) = e^{-\int h(y)dy} \tag{2.19}$$

If $M = yf(xy)$ and $N = xg(xy)$, then

$$I(x, y) = \frac{1}{xM - yN} \tag{2.20}$$

In general, integrating factors are difficult to uncover. If a differential equation does not have one of the forms given above, then a search for an integrating factor likely will not be successful, and other methods of solution are recommended.

(See Problems 2.3–2.6)

Linear Equations

Method of Solution

A first-order *linear* differential equation has the form (see Chapter One)

$$y' + p(x)y = q(x) \tag{2.21}$$

An integrating factor for Equation 2.21 is

$$I(x) = e^{\int p(x)dx} \tag{2.22}$$

Table 2.1

Group of terms	Integrating factor $I(x, y)$	Exact differential $dy(x, y)$
$y\,dx - x\,dy$	$-\dfrac{1}{x^2}$	$\dfrac{x\,dy - y\,dx}{x^2} = d\left(\dfrac{y}{x}\right)$
$y\,dx - x\,dy$	$\dfrac{1}{y^2}$	$\dfrac{y\,dx - x\,dy}{y^2} = d\left(\dfrac{x}{y}\right)$
$y\,dx - x\,dy$	$-\dfrac{1}{xy}$	$\dfrac{x\,dy - y\,dx}{xy} = d\left(\ln\dfrac{y}{x}\right)$
$y\,dx - x\,dy$	$-\dfrac{1}{x^2 + y^2}$	$\dfrac{x\,dy - y\,dx}{x^2 + y^2} = d\left(\arctan\dfrac{y}{x}\right)$
$y\,dx + x\,dy$	$\dfrac{1}{xy}$	$\dfrac{y\,dx + x\,dy}{xy} = d(\ln xy)$
$y\,dx + x\,dy$	$\dfrac{1}{(xy)^n}, \quad n > 1$	$\dfrac{y\,dx + x\,dy}{(xy)^n} = d\left[\dfrac{-1}{(n-1)(xy)^{n-1}}\right]$
$y\,dy + x\,dx$	$\dfrac{1}{x^2 + y^2}$	$\dfrac{y\,dy + x\,dx}{x^2 + y^2} = d\left[\dfrac{1}{2}\ln(x^2 + y^2)\right]$
$y\,dy + x\,dx$	$\dfrac{1}{(x^2 + y^2)^n}, \quad n > 1$	$\dfrac{y\,dy + x\,dx}{(x^2 + y^2)^n} = d\left[\dfrac{-1}{2(n-1)(x^2 + y^2)^{n-1}}\right]$
$ay\,dx + bx\,dy$ (a, b constants)	$x^{a-1}y^{b-1}$	$x^{a-1}y^{b-1}(ay\,dx + bx\,dy) = d(x^a y^b)$

which depends only on x and is independent of y. When both sides of 2.21 are multiplied by $I(x)$, the resulting equation

$$I(x)y' + p(x)I(x)y = I(x)q(x) \qquad (2.23)$$

is exact. This equation can be solved by the method described previously. A simpler procedure is to rewrite 2.23 as

$$\frac{d(yI)}{dx} = Iq(x)$$

integrate both sides of this last equation with respect to x, and then solve the resulting equation for y. The general solution for Equation 2.21 is

$$y = \frac{\int I(x)q(x)dx + c}{I(x)}$$

where c is the constant of integration.
(See Problem 2.7)

Bernoulli Equations

A Bernoulli differential equation has the form

$$y' + p(x)y = q(x)y^n \tag{2.24}$$

where n is a real number. The substitution

$$z = y^{1-n} \tag{2.25}$$

transforms 2.24 into a linear differential equation in the unknown function $z(x)$.
(See Problem 2.8)

Solved Problems

Solved Problem 2.1 Solve $\dfrac{dy}{dx} = \dfrac{x^2 + 2}{y}$.

This equation may be rewritten in the differential form

$$(x^2 + 2)dx - ydy = 0$$

which is separable with $A(x) = x^2 + 2$ and $B(y) = -y$. Its solution is

$$\int (x^2 + 2)dx - \int ydy = c$$

or

$$\frac{1}{3}x^3 + 2x - \frac{1}{2}y^2 = c$$

Solving for y, we obtain the solution in implicit form as

$$y^2 = \frac{2}{3}x^3 + 4x + k$$

with $k = -2c$. Solving for y explicitly, we obtain the two solutions

$$y = \sqrt{\frac{2}{3}x^3 + 4x + k} \quad \text{and} \quad y = -\sqrt{\frac{2}{3}x^3 + 4x + k}$$

Solved Problem 2.2 Solve $y' = \dfrac{y+x}{x}$.

This differential equation is not separable. Instead it has the form $y' = f(x, y)$, with

$$f(x, y) = \frac{y+x}{x}$$

where

$$f(tx, ty) = \frac{ty + tx}{tx} = \frac{t(y+x)}{tx} = \frac{y+x}{x} = f(x, y)$$

so it is homogeneous. Substituting Equations 2.6 and 2.7 into the equation, we obtain

$$v + x\frac{dv}{dx} = \frac{xv + x}{x}$$

which can be algebraically simplified to

$$x\frac{dv}{dx} = 1 \quad \text{or} \quad \frac{1}{x}dx - dv = 0$$

This last equation is separable; its solution is

$$\int \frac{1}{x}dx - \int dv = c$$

which, when evaluated, yields $v = \ln|x| - c$, or

$$v = \ln|kx| \tag{2.26}$$

where we have set $c = -\ln|k|$ and have noted that $\ln|x| + \ln|k| = \ln|xk|$. Finally, substituting $v = y/x$ back into 2.26, we obtain the solution to the given differential equation as $y = x\ln|kx|$.

Solved Problem 2.3 Solve $2xy dx + (1 + x^2) dy = 0$.

This equation has the form of Equation 2.11 with $M(x, y) = 2xy$ and $N(x, y) = 1 + x^2$. Since $\partial M / \partial y = \partial N / \partial x = 2x$, the differential equation is exact. Because this equation is exact, we now determine a function $g(x, y)$ that satisfies Equations 2.14 and 2.15. Substituting $M(x, y) = 2xy$ into 2.14, we obtain $\partial g / \partial x = 2xy$. Integrating both sides of this equation with respect to x, we find

$$\int \frac{\partial g}{\partial x} dx = \int 2xy dx$$

or

$$g(x, y) = x^2 y + h(y) \tag{2.27}$$

Note that when integrating with respect to x, the constant (*with respect to x*) of integration can depend on y.

We now determine $h(y)$. Differentiating 2.27 with respect to y, we obtain $\partial g / \partial y = x^2 + h'(y)$ Substituting this equation along with $N(x, y) = 1 + x^2$ into 2.15, we have

$$x^2 + h'(y) = 1 + x^2 \quad \text{or} \quad h'(y) = 1$$

Integrating this last equation with respect to y, we obtain $h(y) = y + c_1$ ($c_1 = $ constant). Substituting this expression into 2.27 yields

$$g(x, y) = x^2 y + y + c_1$$

The solution to the differential equation, which is given implicitly by 2.16 as $g(x, y) = c$, is

$$x^2 y + y = c_2 \quad (c_2 = c - c_1)$$

Solving for y explicitly, we obtain the solution as $y = c_2 / (x^2 + 1)$.

Solved Problem 2.4 Determine whether the differential equation $y dx - x dy = 0$ is exact.

This equation has the form of Equation 2.11 with $M(x, y) = y$ and $N(x, y) = -x$. Here

$$\frac{\partial M}{\partial y} = 1 \quad \text{and} \quad \frac{\partial N}{\partial x} = -1$$

which are not equal, so the differential equation is not exact.

Solved Problem 2.5 Determine whether $-1/x^2$ is an integrating factor for the differential equation $ydx - xdy = 0$.

It was shown in Problem 2.4 that the differential equation is not exact. Multiplying it by $-1/x^2$, we obtain

$$\frac{-1}{x^2}(ydx - xdy) = 0 \quad \text{or} \quad \frac{-y}{x^2}dx + \frac{1}{x}dy = 0 \qquad (2.28)$$

Equation 2.28 has the form of Equation 2.11 with $M(x, y) = -y/x^2$ and $N(x, y) = 1/x$. Now

$$\frac{\partial M}{\partial y} = \frac{\partial}{\partial y}\left(\frac{-y}{x^2}\right) = \frac{-1}{x^2} = \frac{\partial}{\partial x}\left(\frac{1}{x}\right) = \frac{\partial N}{\partial x}$$

so 2.28 is exact, which implies that $-1/x^2$ is an integrating factor for the original differential equation.

Solved Problem 2.6 Solve $ydx - xdy = 0$.

Using the results of Problem 2.5, we can rewrite the given differential equation as

$$\frac{xdy - ydx}{x^2} = 0$$

which is exact. Equation 2.28 can be solved using the steps described in Equations 2.14 through 2.16.

Alternatively, we note from Table 2.1 that 2.28 can be rewritten as $d(y/x) = 0$. Hence, by direct integration, we have $y/x = c$, or $y = xc$, as the solution.

Solved Problem 2.7 Solve $y' + (4/x)y = x^4$.

The differential equation has the form of Equation 2.21, with $p(x) = 4/x$ and $q(x) = x^4$, and is linear. Here

$$\int p(x)dx = \int \frac{4}{x}dx = 4\ln |x| = \ln x^4$$

so 2.22 becomes

$$I(x) = e^{\int p(x)dx} = e^{\ln x^4} = x^4 \qquad (2.29)$$

Multiplying the differential equation by the integrating factor defined by 2.29, we obtain

$$x^4 y' + 4x^3 y = x^8 \quad \text{or} \quad \frac{d}{dx}(yx^4) = x^8$$

Integrating both sides of this last equation with respect to x, we obtain

$$yx^4 = \frac{1}{9}x^9 + c \quad \text{or} \quad y = \frac{c}{x^4} + \frac{1}{9}x^5$$

Solved Problem 2.8 Solve $y' + xy = xy^2$.

This equation is not linear. It is, however, a Bernoulli differential equation having the form of Equation 2.24 with $p(x) = q(x) = x$, and $n = 2$. We make the substitution suggested by 2.25, namely $z = y^{1-2} = y^{-1}$, from which follow

$$y = \frac{1}{z} \quad \text{and} \quad y' = -\frac{z'}{z^2}$$

Substituting these equations into the differential equation, we obtain

$$-\frac{z'}{z^2} + \frac{x}{z} = \frac{x}{z^2} \quad \text{or} \quad z' - xz = -x$$

This last equation is linear for the unknown function $z(x)$. It has the form of Equation 2.21 with y replaced by z and $p(x) = q(x) = -x$. The integrating factor is

$$I(x) = e^{\int (-x)dx} = e^{-x^2/2}$$

Multiplying the differential equation by $I(x)$, we obtain

$$e^{-x^2/2} \frac{dz}{dx} - xe^{-x^2/2}z = -xe^{-x^2/2}$$

or

$$\frac{d}{dx}\left(ze^{-x^2/2}\right) = -xe^{-x^2/2}$$

Upon integrating both sides of this last equation, we have

$$ze^{-x^2/2} = e^{-x^2/2} + c$$

whereupon

$$z(x) = ce^{x^2/2} + 1$$

The solution of the original differential equation is then

$$y = \frac{1}{z} = \frac{1}{ce^{x^2/2} + 1}$$

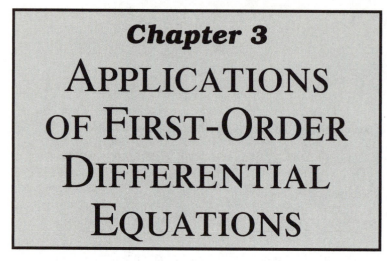

Chapter 3
APPLICATIONS OF FIRST-ORDER DIFFERENTIAL EQUATIONS

IN THIS CHAPTER:

- ✔ *Growth and Decay Problems*
- ✔ *Temperature Problems*
- ✔ *Falling Body Problems*
- ✔ *Dilution Problems*
- ✔ *Electrical Circuits*
- ✔ *Orthogonal Trajectories*
- ✔ *Solved Problems*

Growth and Decay Problems

Let $N(t)$ denote the amount of substance (or population) that is either growing or decaying. If we assume that dN/dt, the time rate of change of this amount of substance, is proportional to the amount of substance present, then $dN/dt = kN$, or

$$\frac{dN}{dt} - kN = 0 \tag{3.1}$$

where k is the constant of proportionality.

We are assuming that $N(t)$ is a differentiable, hence continuous, function of time. For population problems, where $N(t)$ is actually discrete and integer-valued, this assumption is incorrect. Nonetheless, 3.1 still provides a good approximation to the physical laws governing such a system.

Temperature Problems

Newton's law of cooling, which is equally applicable to heating, states that *the time rate of change of the temperature of a body is proportional to the temperature difference between the body and its surrounding medium.* Let T denote the temperature of the body and let T_m denote the temperature of the surrounding medium. Then the time rate of change of the temperature of the body is dT/dt, and Newton's law of cooling can be formulated as $dT/dt = -k(T - T_m)$, or as

$$\frac{dT}{dt} + kT = kT_m \tag{3.2}$$

where k is a *positive* constant of proportionality. Once k is chosen positive, the minus sign is required in Newton's law to make dT/dt negative in a cooling process, when T is greater than T_m, and positive in a heating process, when T is less than T_m.

Falling Body Problems

Consider a vertically falling body of mass m that is being influenced only by gravity g and an air resistance that is proportional to the velocity of the body. Assume that both gravity and mass remain constant and, for convenience, choose the downward direction as the positive direction.

You Need to Know ✔

Newton's second law of motion: *The net force acting on a body is equal to the time rate of change of the momentum of the body; or, for constant mass,*

$$F = m\frac{dv}{dt} \tag{3.3}$$

where F is the net force on the body and v is the velocity of the body, both at time t.

For the problem at hand, there are two forces acting on the body: the force due to gravity given by the weight w of the body, which equals mg, and the force due to air resistance given by $-kv$, where $k \geq 0$ is a constant of proportionality. The minus sign is required because this force opposes the velocity; that is, it acts in the upward, or negative, direction (see Figure 3-1). The net force F on the body is, therefore, $F = mg - kv$. Substituting this result into 3.3, we obtain

$$mg - kv = m\frac{dv}{dt}$$

or

$$\frac{dv}{dt} + \frac{k}{m}v = g \tag{3.4}$$

as the equation of motion for the body.

If air resistance is negligible or nonexistent, then $k = 0$ and 3.4 simplifies to

$$\frac{dv}{dt} = g \tag{3.5}$$

When $k > 0$, the limiting velocity v_l is defined by

$$v_l = \frac{mg}{k} \tag{3.6}$$

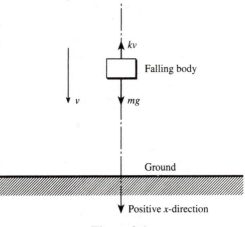

Figure 3-1

Caution: Equations 3.4, 3.5, and 3.6 are valid only if the given conditions are satisfied. These equations are not valid if, for example, air resistance is not proportional to velocity but to the velocity squared, or if the upward direction is taken to be the positive direction.

Dilution Problems

Consider a tank which initially holds V_0 gal of brine that contains a lb of salt. Another solution, containing b lb of salt per gallon, is poured into the tank at the rate of e gal/min while simultaneously, the well-stirred solution leaves the tank at the rate of f gal/min (Figure 3-2). The problem is to find the amount of salt in the tank at any time t.

Let Q denote the amount (in pounds) of salt in the tank at any time. The time rate of change of Q, dQ/dt, equals the rate at which salt enters the tank minus the rate at which salt leaves the tank. Salt enters the tank at the rate of be lb/min. To determine the rate at which salt leaves the tank, we first calculate the volume of brine in the tank at any time t, which is the initial volume V_0 plus the volume of brine added et minus the volume of brine removed ft. Thus, the volume of brine at any time is

$$V_0 + et - ft \tag{3.7}$$

The concentration of salt in the tank at any time is $Q/(V_0 + et - ft)$, from which it follows that salt leaves the tank at the rate of

$$f\left(\frac{Q}{V_0 + et - ft}\right) \text{ lb/min}$$

Thus,

$$\frac{dQ}{dt} = be - f\left(\frac{Q}{V_0 + et - ft}\right)$$

or

$$\frac{dQ}{dt} + \frac{f}{V_0 + (e - f)t}Q = be \tag{3.8}$$

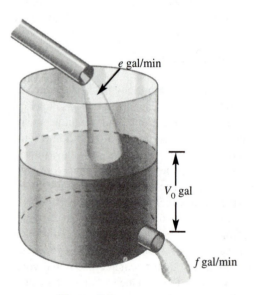

e gal/min

V_0 gal

f gal/min

Figure 3-2

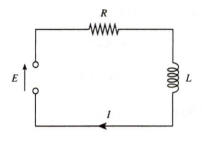

Figure 3-3

Electrical Circuits

The basic equation governing the amount of current I (in amperes) in a simple RL circuit (see Figure 3-3) consisting of a resistance R (in ohms), an inductor L (in henries), and an electromotive force (abbreviated emf) E (in volts) is

$$\frac{dI}{dt} + \frac{R}{L}I = \frac{E}{L} \qquad (3.9)$$

For an RC circuit consisting of a resistance, a capacitance C (in farads), an emf, and no inductance (Figure 3-4), the equation governing the amount of electrical charge q (in coulombs) on the capacitor is

$$\frac{dq}{dt} + \frac{1}{RC}q = \frac{E}{R} \qquad (3.10)$$

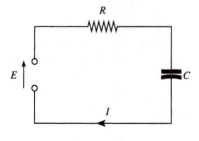

Figure 3-4

The relationship between q and I is

$$I = \frac{dq}{dt} \tag{3.11}$$

For more complex circuits see Chapter Seven.

Orthogonal Trajectories

Consider a one-parameter family of curves in the xy-plane defined by

$$F(x, y, c) = 0 \tag{3.12}$$

where c denotes the parameter. The problem is to find another one-parameter family of curves, called the *orthogonal trajectories* of the family of curves in 3.12 and given analytically by

$$G(x, y, k) = 0 \tag{3.13}$$

such that every curve in this new family 3.13 intersects at right angles every curve in the original family 3.12.

We first implicitly differentiate 3.12 with respect to x, then eliminate c between this derived equation and 3.12. This gives an equation connecting x, y, and y', which we solve for y' to obtain a differential equation of the form

$$\frac{dy}{dx} = f(x, y) \tag{3.14}$$

The orthogonal trajectories of 3.12 are the solutions of

$$\frac{dy}{dx} = -\frac{1}{f(x, y)} \tag{3.15}$$

For many families of curves, one cannot explicitly solve for dy/dx and obtain a differential equation of the form 3.14. We do not consider such curves in this book.

Solved Problems

Solved Problem 3.1 A bacteria culture is known to grow at a rate proportional to the amount present. After one hour, 1000 strands of the bacteria are observed in the culture; and after four hours, 3000 strands. Find (*a*) an expression for the approximate number of strands of the bacteria present in the culture at any time *t* and (*b*) the approximate number of strands of the bacteria originally in the culture.

(*a*) Let $N(t)$ denote the number of bacteria strands in the culture at time *t*. From Equation 3.1, $dN/dt - kN = 0$, which is both linear and separable. Its solution is

$$N(t) = ce^{kt} \tag{3.16}$$

At $t = 1$, $N = 1000$; hence,

$$1000 = ce^k \tag{3.17}$$

At $t = 4$, N = *3000*; hence.

$$3000 = ce^{4k} \tag{3.18}$$

Solving 3.17 and 3.18 for *k* and *c*, we find

$$k = \frac{1}{3}\ln 3 \approx 0.3662 \text{ and } c = 1000e^{-k} = 693$$

Substituting these values of *k* and *c* into 3.16, we obtain

$$N(t) = 693e^{0.3662t} \tag{3.19}$$

(*b*) *We require N* at $t = 0$. Substituting $t = 0$ into 3.19, we obtain $N(0)$ $= 693e^{(0.3662)(0)} = 693$.

Solved Problem 3.2 A tank initially holds 100 gal of a brine solution containing 20 lb of salt. At $t = 0$, fresh water is poured into the tank at the rate of 5 gal/min, while the well-stirred mixture leaves the tank at the same rate. Find the amount of salt in the tank at any time *t*.

Here, $V_0 = 100$, $a = 20$, $b = 0$, and $e = f = 5$. Equation 3.8 becomes

$$\frac{dQ}{dt} + \frac{1}{20}Q = 0$$

The solution of this linear equation is

$$Q = ce^{-t/20} \qquad\qquad (3.20)$$

At $t = 0$, we are given that $Q = a = 20$. Substituting these values into 3.20, we find that $c = 20$, so that 3.20 can be rewritten as $Q = 20e^{-t/20}$. Note that as $t \to \infty$, $Q \to 0$ as it should, since only fresh water is being added.

Chapter 4
LINEAR DIFFERENTIAL EQUATIONS: THEORY OF SOLUTIONS

IN THIS CHAPTER:

- ✔ *Linear Differential Equations*
- ✔ *Linearly Independent Solutions*
- ✔ *The Wronskian*
- ✔ *Nonhomogeneous Equations*

Linear Differential Equations

An nth-order linear differential equation has the form

$$b_n(x)y^{(n)} + b_{n-1}(x)y^{(n-1)} + \cdots + b_1(x)y' + b_0(x)y = g(x) \qquad (4.1)$$

where $g(x)$ and the coefficients $b_j(x)$ ($j = 0,1,2,...,n$) depend solely on the variable x. In other words, they do *not* depend on y or any derivative of y.

If $g(x) = 0$, then Equation 4.1 is *homogeneous*; if not, 4.1 is *nonhomogeneous*. A linear differential equation has *constant coefficients* if all the coefficients $b_j(x)$ in 4.1 are constants; if one or more of these coefficients is not constant, 4.1 has *variable coefficients*.

Theorem 4.1. Consider the initial-value problem given by the linear differential equation 4.1 and the n initial conditions

$$y(x_0) = c_0, \quad y'(x_0) = c_1,$$
$$y''(x_0) = c_2, \ldots, y^{(n-1)}(x_0) = c_{n-1} \tag{4.2}$$

If $g(x)$ and $b_j(x)$ ($j = 0,1,2,\ldots, n$) are continuous in some interval $\mathcal{I}$ containing x_0 and if $b_n(x) \neq 0$ in $\mathcal{I}$, then the initial-value problem given by 4.1 and 4.2 has a unique (only one) solution defined throughout $\mathcal{I}$.

When the conditions on $b_n(x)$ in Theorem 4.1 hold, we can divide Equation 4.1 by $b_n(x)$ to get

$$y^{(n)} + a_{n-1}(x)y^{(n-1)} + \cdots + a_2(x)y'' + a_1(x)y' + a_0(x)y = \phi(x) \tag{4.3}$$

where $a_j(x) = b_j(x)/b_n(x)$ ($j = 0,1,2,\ldots, n-1$) and $\phi(x) = g(x)/b_n(x)$.
Let us define the differential operator $\mathbf{L}(y)$ by

$$\mathbf{L}(y) \equiv y^{(n)} + a_{n-1}(x)y^{(n-1)} + \cdots + a_2(x)y'' + a_1(x)y' + a_0(x)y \tag{4.4}$$

where $a_i(x)$ ($i = 0,1,2,\ldots, n-1$) is continuous on some interval of interest. Then 4.3 can be rewritten as

$$\mathbf{L}(y) = \phi(x) \tag{4.5}$$

and, in particular, a linear *homogeneous* differential equation can be expressed as

$$\mathbf{L}(y) = 0 \tag{4.6}$$

Linearly Independent Solutions

A set of functions $\{y_1(x), y_2(x),...,y_n(x)\}$ is *linearly dependent* on $a \leq x \leq b$ if there exist constants $c_1, c_2,...,c_n$ *not all zero*, such that

$$c_1 y_1(x) + c_2 y_2(x) + \cdots + c_n y_n(x) \equiv 0 \qquad (4.7)$$

on $a \leq x \leq b$.

Example 4.1: The set $\{x, 5x, 1, \sin x\}$ is linearly dependent on $[-1,1]$ since there exist constants $c_1 = -5$, $c_2 = 1$, $c_3 = 0$, and $c_4 = 0$, *not all zero*, such that 4.7 is satisfied. In particular,

$$-5 \cdot x + 1 \cdot 5x + 0 \cdot 1 + 0 \cdot \sin x = 0$$

Note that $c_1 = c_2 = \cdots c_n = 0$ is a set of constants that always satisfies 4.7. A set of functions is linearly dependent if there exists *another* set of constants, *not all zero*, that also satisfies 4.7. If the *only* solution to 4.7 is $c_1 = c_2 = \cdots c_n = 0$, then the set of functions $\{y_1(x), y_2(x),...,y_n(x)\}$ is *linearly independent* on $a \leq x \leq b$.

Theorem 4.2. The *nth*-order linear *homogeneous* differential equation $\mathbf{L}(y) = 0$ always has n linearly independent solutions. If $y_1(x), y_2(x),...,y_n(x)$ represent these solutions, then the general solution of $\mathbf{L}(y) = 0$ is

$$y(x) = c_1 y_1(x) + c_2 y_2(x) + \cdots + c_n y_n(x) \qquad (4.8)$$

where $c_1, c_2,...,c_n$ denote arbitrary constants.

The Wronskian

The *Wronskian* of a set of functions $\{z_1(x), z_2(x),..., z_n(x)\}$ on the interval $a \leq x \leq b$, having the property that each function possesses $n - 1$ derivatives on this interval, is the determinant

$$W(z_1, z_2,..., z_n) = \begin{vmatrix} z_1 & z_2 & \cdots & z_n \\ z_1' & z_2' & \cdots & z_n' \\ z_1'' & z_2'' & \cdots & z_n'' \\ \vdots & \vdots & \cdots & \vdots \\ z_1^{(n-1)} & z_2^{(n-1)} & \cdots & z_n^{(n-1)} \end{vmatrix}$$

Theorem 4.3. If the Wronskian of a set of n functions defined on the interval $a \leq x \leq b$ is nonzero for at least one point in this interval, then the set of functions is linearly independent there. If the Wronskian is identically zero on this interval and if each of the functions is a solution to the same linear differential equation, then the set of functions is linearly dependent.

Caution: Theorem 4.3 is silent when the Wronskian is identically zero and the functions are not known to be solutions of the same linear differential equation. In this case, one must test directly whether Equation 4.7 is satisfied.

Nonhomogeneous Equations

Let y_p denote any *particular* solution of Equation 4.5 (see Chapter One) and let y_h (henceforth called the *homogeneous* or *complementary solution*) represent the *general* solution of the associated homogeneous equation $\mathbf{L}(y) = 0$.

Theorem 4.4. The general solution to $\mathbf{L}(y) = \phi(x)$ is

$$y = y_h + y_p \tag{4.9}$$

Don't Forget

The general solution is the sum of the homogeneous and particular solutions.

Chapter 5

SOLUTIONS OF LINEAR HOMOGENEOUS DIFFERENTIAL EQUATIONS WITH CONSTANT COEFFICIENTS

IN THIS CHAPTER:

- ✔ The Characteristic Equation
- ✔ General Solution for Second-Order Equations
- ✔ General Solution for nth-Order Equations
- ✔ Solved Problems

The Characteristic Equation

Second-Order Equations

Corresponding to the differential equation

$$y'' + a_1 y' + a_0 y = 0 \tag{5.1}$$

in which a_1 and a_0 are constants, is the algebraic equation

$$\lambda^2 + a_1\lambda + a_0 = 0 \tag{5.2}$$

which is obtained by substituting $y = e^{\lambda x}$ (assuming x to be the independent variable) into Equation 5.1 and simplifying. Note that Equation 5.2 can be easily obtained by replacing y'', y', and y by λ^2, λ^1, and, $\lambda^0 = 1$, respectively. Equation 5.2 is called the *characteristic equation* of 5.1.

Example 5.1 The characteristic equation of $y'' + 3y' - 4y = 0$ is $\lambda^2 + 3\lambda - 4 = 0$; the characteristic equation of $y'' - 2y' + y = 0$ is $\lambda^2 - 2\lambda + 1 = 0$.

Characteristic equations for differential equations having dependent variables other than y are obtained analogously, by replacing the jth derivative of the dependent variable by λ^j ($j = 0,1,2$).

nth-Order Equations

Similarly, the characteristic equation of the differential equation

$$y^{(n)} + a_{n-1}y^{(n-1)} + \cdots + a_1 y' + a_0 y = 0 \tag{5.3}$$

with constant coefficients a_j ($j = 0,1,..., n-1$) is

$$\lambda^n + a_{n-1}\lambda^{n-1} + \cdots + a_1\lambda + a_0 = 0 \tag{5.4}$$

The characteristic equation 5.4 is obtained from 5.3 by replacing $y^{(j)}$ by λ^j ($j = 0,1,..., n$). Characteristic equations for differential equations having dependent variables other than y are obtained analogously, by replacing the jth derivative of the dependent variable by λ^j ($j = 0,1,..., n$).

Example 5.2. The characteristic equation of $y^{(4)} - 3y''' + 2y'' - y = 0$ is $\lambda^4 - 3\lambda^3 + 2\lambda^2 - 1 = 0$. The characteristic equation of

$$\frac{d^5x}{dt^5} - 3\frac{d^3x}{dt^3} + 5\frac{dx}{dt} - 7x = 0$$

is

$$\lambda^5 - 3\lambda^3 + 5\lambda - 7 = 0$$

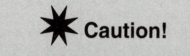 **Caution!**

Characteristic equations are only defined for linear homogeneous differential equations with constant coefficients.

General Solution for Second-Order Equations

The characteristic equation (5.2) can be factored into

$$(\lambda - \lambda_1)(\lambda - \lambda_2) = 0 \qquad (5.5)$$

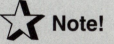 **Note!**

The roots of the characteristic polynomial determine the solution of the differential equation.

There are three cases to consider.

Case 1. λ_1 **and** λ_2 **both real and distinct.** Two linearly independent solutions are $e^{\lambda_1 x}$ and $e^{\lambda_2 x}$, and the general solution is (Theorem 4.2)

$$y = c_1 e^{\lambda_1 x} + c_2 e^{\lambda_2 x} \tag{5.6}$$

In the special case $\lambda_2 = -\lambda_1$, the solution 5.6 can be rewritten as $y = k_1 \cosh \lambda_1 x + k_2 \sinh \lambda_1 x$.

Case 2. $\lambda_1 = a + ib$, **a complex number.** Since a_1 and a_0 in 5.1 and 5.2 are assumed real, the roots of 5.2 must appear in conjugate pairs; thus, the other root is $\lambda_2 = a - ib$. Two linearly independent solutions are $e^{(a+ib)x}$, and $e^{(a-ib)x}$, and the general complex solution is

$$y = d_1 e^{(a+ib)x} + d_2 e^{(a-ib)x} \tag{5.7}$$

which is algebraically equivalent to

$$y = c_1 e^{ax} \cos bx + c_2 e^{ax} \sin bx \tag{5.8}$$

Case 3. $\lambda_1 = \lambda_2$. Two linearly independent solutions are $e^{\lambda_1 x}$ and $xe^{\lambda_1 x}$, and the general solution is

$$y = c_1 e^{\lambda_1 x} + c_2 x e^{\lambda_1 x} \tag{5.9}$$

Warning: The above solutions are *not valid* if the differential equation is not linear or does not have constant coefficients. Consider, for example, the equation $y'' - x^2 y = 0$. The roots of the characteristic equation are $\lambda_1 = x$ and $\lambda_2 = -x$, but the solution *is not*

$$y = c_1 e^{(x)x} + c_2 e^{(-x)x} = c_1 e^{x^2} + c_2 e^{-x^2}$$

Linear equations with variable coefficients are considered in Chapter Twelve.

General Solution for *n*th-Order Equations

The general solution of 5.3 is obtained directly from the roots of 5.4. If the roots $\lambda_1, \lambda_2,..., \lambda_n$ are all real and distinct, the solution is

$$y = c_1 e^{\lambda_1 x} + c_2 e^{\lambda_2 x} + \cdots + c_n e^{\lambda_n x} \tag{5.10}$$

If the roots are distinct, but some are complex, then the solution is again given by 5.10. As in the second-order equation, those terms involving complex exponentials can be combined to yield terms involving sines and cosines. If λ_k is a root of multiplicity p [that is, if $(\lambda - \lambda_k)^p$ is a factor of the characteristic equation, but $(\lambda - \lambda_k)^{p+1}$ is not] then there will be p linearly independent solutions associated  with λ_k given by $e^{\lambda_k x}, xe^{\lambda_k x}, x^2 e^{\lambda_k x}, \ldots, x^{p-1} e^{\lambda_k x}$. These solutions are combined in the usual way with the solutions associated with the other roots to obtain the complete solution.

In theory it is always possible to factor the characteristic equation, but in practice this can be extremely difficult, especially for differential equations of high order. In such cases, one must often use numerical techniques to approximate the solutions. See Chapter Fourteen.

Solved Problems

Solved Problem 5.1 Solve $y'' - y' - 2y = 0$.

The characteristic equation is $\lambda^2 - \lambda - 2 = 0$, which can be factored into $(\lambda - 2)(\lambda + 1) = 0$. Since the roots $\lambda_1 = 2$ and $\lambda_2 = -1$ are real and distinct, the solution is given by 5.6 as

$$y = c_1 e^{2x} + c_2 e^{-x}$$

Solved Problem 5.2 Solve $y'' - 8y' + 16y = 0$.

The characteristic equation is $\lambda^2 - 8\lambda + 16 = 0$ which can be factored into $(\lambda - 4)^2 = 0$. The roots $\lambda_1 = \lambda_2 = 4$ are real and equal, so the general solution is given by 5.9 as

$$y = c_1 e^{4x} + c_2 x e^{4x}$$

Solved Problem 5.3 Solve $y''' - 6y'' + 2y' + 36y = 0$.

The characteristic equation $\lambda^3 - 6\lambda^2 + 2\lambda + 36 = 0$, has roots $\lambda_1 = -2$, $\lambda_2 = 4 + i\sqrt{2}$, and $\lambda_3 = 4 - i\sqrt{2}$. The solution is

$$y = c_1 e^{-2x} + d_2 e^{(4+i\sqrt{2})x} + d_3 e^{(4-i\sqrt{2})x}$$

which can be rewritten, using Euler's relations

$$e^{ibx} = \cos bx + i\sin bx \quad \text{and} \quad e^{-ibx} = \cos bx - i\sin bx$$

as

$$y = c_1 e^{-2x} + d_2 e^{4x} e^{i\sqrt{2}x} + d_3 e^{4x} e^{-i\sqrt{2}x}$$
$$y = c_1 e^{-2x} + d_2 e^{4x}(\cos\sqrt{2}x + i\sin\sqrt{2}x) + d_3 e^{4x}(\cos\sqrt{2}x - i\sin\sqrt{2}x)$$
$$y = c_1 e^{-2x} + (d_2 + d_3)e^{4x}\cos\sqrt{2}x + i(d_2 - d_3)e^{4x}\sin\sqrt{2}x$$
$$y = c_1 e^{-2x} + c_2 e^{4x}\cos\sqrt{2}x + c_3 e^{4x}\sin\sqrt{2}x$$

Note that this form of the solution corresponding to the complex roots can be easily formulated using Equation 5.8.

Chapter 6

SOLUTIONS OF LINEAR NONHOMOGENEOUS EQUATIONS AND INITIAL-VALUE PROBLEMS

IN THIS CHAPTER:

- ✔ *The Method of Undetermined Coefficients*
- ✔ *Variation of Parameters*
- ✔ *Initial-Value Problems*
- ✔ *Solved Problems*

The general solution to the linear differential equation $\mathbf{L}(y) = \phi(x)$ is given by Theorem 4.4 as $y = y_h + y_p$ where y_p denotes one solution to the differential equation and y_h is the general solution to the associated homogeneous equation, $\mathbf{L}(y) = 0$. Methods for obtaining y_h when the differential equation has constant coefficients are given in Chapter Five. In this chapter, we give methods for obtaining a particular solution y_p *once* y_h *is known.*

The Method of Undetermined Coefficients

Simple Form of the Method

The *method of undetermined coefficients* is applicable only if $\phi(x)$ and *all* of its derivatives can be written in terms of the same *finite* set of linearly independent functions, which we denote by $\{y_1(x), y_2(x),..., y_n(x)\}$. The method is initiated by assuming a particular solution of the form

$$y_p(x) = A_1 y_1(x) + A_2 y_2(x) + \cdots + A_n y_n(x)$$

where $A_1, A_2,..., A_n$ denote arbitrary multiplicative constants. These arbitrary constants are then evaluated by substituting the proposed solution into the given differential equation and equating the coefficients of like terms.

Case 1. $\phi(x) = p_n(x)$, **an nth-degree polynomial in x.** Assume a solution of the form

$$y_p = e^{\alpha x}(A_n x^n + A_{n-1} x^{n-1} + \cdots + A_1 x - \tag{6.1}$$

where A_j ($j = 0,1,2,..., n$) is a constant to be determined.

Case 2. $\phi(x) = ke^{\alpha x}$ **where k and α are known constants.** Assume a solution of the form

$$y_p = Ae^{\alpha x} \tag{6.2}$$

where A is a constant to be determined.

Case 3. $\phi(x) = k_1 \sin \beta x + k_2 \cos \beta x$ **where k_1, k_2, and β are known constants.** Assume a solution of the form

$$y_p = A \sin \beta x + B \cos \beta x \tag{6.3}$$

where A and B are constants to be determined.

Don't Forget

$y_p = A \sin \beta x + B \cos \beta x$ in its entirety is assumed for $\phi(x) = k_1 \sin \beta x + k_2 \cos \beta x$ even when k_1 or k_2 is zero, because the derivatives of sines or cosines involve both sines and cosines.

Generalizations

If $\phi(x)$ is the product of terms considered in Cases 1 through 3, take y_p to be the product of the corresponding assumed solutions and algebraically combine arbitrary constants where possible. In particular, if $\phi(x) = e^{\alpha x} p_n(x)$ is the product of a polynomial with an exponential, assume

$$y_p = e^{\alpha x}(A_n x^n + A_{n-1} x^{n-1} + \cdots + A_1 x + A_0) \tag{6.4}$$

where A_j is as in Case 1. If, instead, $\phi(x) = e^{\alpha x} p_n(x) \sin \beta x$ is the product of a polynomial, exponential, and sine term, or if $\phi(x) = e^{\alpha x} p_n(x) \cos \beta x$ is the product of a polynomial, exponential, and cosine term, then assume

$$\begin{aligned} y_p = \ & e^{\alpha x} \sin \beta x (A_n x^n + A_{n-1} x^{n-1} + \cdots + A_1 x + A_0) \\ &+ e^{\alpha x} \cos \beta x (B_n x^n + B_{n-1} x^{n-1} + \cdots + B_1 x + B_0) \end{aligned} \tag{6.5}$$

where A_j and B_j ($j = 0,1,2,..., n$) are constants which still must be determined.

If $\phi(x)$ is the sum (or difference) of terms already considered, then we take y_p to be the sum (or difference) of the corresponding assumed solutions and algebraically combine arbitrary constants where possible.

Modifications

If any term of the assumed solution, disregarding multiplicative constants, is also a term of y_h (the homogeneous solution), then the assumed solution must be modified by multiplying it by x^m, where m is the smallest positive integer such that the product of x^m with the assumed solution has no terms in common with y_h.

Limitations of the Method

In general, if $\phi(x)$ is not one of the types of functions considered above, or if the differential equation *does not have constant coefficients*, then the following method is preferred.

Variation of Parameters

Variation of parameters is another method for finding a particular solution of the nth-order linear differential equation

$$L(y) = \phi(x) \tag{6.6}$$

once the solution of the associated homogeneous equation $L(y) = 0$ is known. Recall from Theorem 4.2 that if $y_1(x)$, $y_2(x)$,..., $y_n(x)$ are n linearly independent solutions of $L(y) = 0$, then the general solution of $L(y) = 0$ is

$$y_h = c_1 y_1(x) + c_2 y_2(x) + \cdots + c_n y_n(x) \tag{6.7}$$

The Method

A particular solution of $L(y) = \phi(x)$ has the form

$$y_p = v_1 y_1 + v_2 y_2 + \cdots + v_n y_n \tag{6.8}$$

where $y_i = y_i(x)$ $(i = 1,2,..., n)$ is given in Equation 6.7 and v_i $(i = 1,2,..., n)$ is an unknown function of x which still must be determined.

To find v_i, first solve the following linear equations simultaneously for v_i':

$$v_1'y_1 + v_2'y_2 + \cdots + v_n'y_n = 0$$
$$v_1'y_1' + v_2'y_2' + \cdots + v_n'y_n' = 0$$
$$\vdots$$
$$v_1'y_1^{(n-2)} + v_2'y_2^{(n-2)} + \cdots + v_n'y_n^{(n-2)} = 0$$
$$v_1'y_1^{(n-1)} + v_2'y_2^{(n-1)} + \cdots + v_n'y_n^{(n-1)} = \phi(x)$$

(6.9)

Then integrate each v_i' to obtain v_i, disregarding all constants of integration. This is permissible because we are seeking only *one* particular solution.

Example 6.1: For the special case $n = 3$, Equations 6.9 reduce to

$$v_1'y_1 + v_2'y_2 + v_3'y_3 = 0$$
$$v_1'y_1' + v_2'y_2' + v_3'y_3' = 0$$
$$v_1'y_1'' + v_2'y_2'' + v_3'y_3'' = \phi(x)$$

(6.10)

For the case $n = 2$, Equations 6.9 become

$$v_1'y_1 + v_2'y_2 = 0$$
$$v_1'y_1' + v_2'y_2' = \phi(x)$$

(6.11)

and for the case $n = 1$, we obtain the single equation

$$v_1'y_1 = \phi(x)$$

(6.12)

Since $y_1(x), y_2(x),..., y_n(x)$ are n linearly independent solutions of the same equation $L(y) = 0$, their Wronskian is not zero (Theorem 4.3). This means that the system 6.9 has a nonzero determinant and can be solved uniquely for $v_1'(x), v_2'(x),..., v_n'(x)$.

Scope of the Method

The method of variation of parameters can be applied to *all* linear differential equations. It is therefore more powerful than the method of undetermined coefficients, which is restricted to linear differential equations with constant coefficients and particular forms of $\phi(x)$. Nonetheless, in those cases where both methods are applicable, the method of undetermined coefficients is usually the more efficient and, hence, preferable. As a practical matter, the integration of $v_i'(x)$ may be impossible to perform. In such an event other methods (in particular, numerical techniques) must be employed.

Initial-Value Problems

Initial-value problems are solved by applying the initial conditions to the general solution of the differential equation. It must be emphasized that the initial conditions are applied *only* to the general solution and *not* to the homogeneous solution y_h that possesses all the arbitrary constants that must be evaluated. The one exception is when the general solution is the homogeneous solution; that is, when the differential equation under consideration is itself homogeneous.

Solved Problems

Solved Problem 6.1 Solve $y'' - y' - 2y = 4x^2$.

From Problem 5.1, $y_h = c_1 e^{2x} + c_2 e^{-x}$. Here $\phi(x) = 4x^2$, a second degree polynomial. Using Equation 6.1, we assume that

$$y_p = A_2 x^2 + A_1 x + A_0 \qquad (6.13)$$

Thus, $y_p' = 2A_2 x + A_1$ and $y_p'' = 2A_2$. Substituting these results into the differential equation, we have

$$2A_2 - (2A_2 x + A_1) - 2(A_2 x^2 + A_1 x + A_0) = 4x^2$$

or, equivalently,

$$(-2A_2)x^2 + (-2A_2 - 2A_1)x + (2A_2 - A_1 - 2A_0) = 4x^2 + (0)x + 0$$

Equating the coefficients of like powers of x, we obtain

$$-2A_2 = 4 \quad -2A_2 - 2A_1 = 0 \quad 2A_2 - A_1 - 2A_0 = 0$$

Solving this system, we find that $A_2 = -2$, $A_1 = 2$, and $A_0 = -3$. Hence Equation 6.13 becomes

$$y_p = -2x^2 + 2x - 3$$

and the general solution is

$$y = y_h + y_p = c_1 e^{2x} + c_2 e^{-x} - 2x^2 + 2x - 3$$

Solved Problem 6.2 Solve $y'' - y' - 2y = \sin 2x$.

Again by Problem 5.1, $y_h = c_1 e^{2x} + c_2 e^{-x}$. Here $\phi(x)$ has the form displayed in Case 3 with $k_1 = 1$, $k_2 = 0$, and $\beta = 2$. Using Equation 6.3, we assume that

$$y_p = A\sin 2x + B\cos 2x \qquad (6.14)$$

Thus, $y_p' = 2A\cos 2x - 2B\sin 2x$ and $y_p'' = -4A\sin 2x - 4B\cos 2x$. Substituting these results into the differential equation, we have

$$(-4A\sin 2x - 4B\cos 2x) - (2A\cos 2x - 2B\sin 2x)$$
$$-2(A\sin 2x + B\cos 2x) = \sin 2x$$

or, equivalently,

$$(-6A + 2B)\sin 2x + (-6B - 2A)\cos 2x = (1)\sin 2x + (0)\cos 2x$$

Equating coefficients of like terms, we obtain

$$-6A + 2B = 1 \qquad -2A - 6B = 0$$

Solving this system, we find that $A = -3/20$ and $B = 1/20$. Then from Equation 6.14,

$$y_p = -\frac{3}{20}\sin 2x + \frac{1}{20}\cos 2x$$

and the general solution is

$$y = c_1 e^{2x} + c_2 e^{-x} - \frac{3}{20}\sin 2x + \frac{1}{20}\cos 2x$$

Solved Problem 6.3 Solve $y''' + y' = \sec x$.

This is a third-order equation with

$$y_h = c_1 + c_2 \cos x + c_3 \sin x$$

It follows from Equation 6.8 that

$$y_p = v_1 + v_2 \cos x + v_3 \sin x \tag{6.15}$$

Here $y_1 = 1$, $y_2 = \cos x$, $y_3 = \sin x$, and $\phi(x) = \sec x$, so Equation 6.10 becomes

$$v_1'(1) + v_2'(\cos x) + v_3'(\sin x) = 0$$
$$v_1'(0) + v_2'(-\sin x) + v_3'(\cos x) = 0$$
$$v_1'(0) + v_2'(-\cos x) + v_3'(-\sin x) = \sec x$$

Solving this set of equations simultaneously, we obtain $v_1' = \sec x$, $v_2' = -1$, and $v_3' = -\tan x$. Thus,

$$v_1 = \int v_1' dx = \int \sec x \, dx = \ln|\sec x + \tan x|$$
$$v_2 = \int v_2' dx = \int -1 dx = -x$$
$$v_3 = \int v_3' dx = \int -\tan x \, dx = -\int \frac{\sin x}{\cos x} dx = \ln|\cos x|$$

Substituting these values into Equation 6.15, we obtain

$$y_p = \ln|\sec x + \tan x| - x\cos x + (\sin x)\ln|\cos x|$$

The general solution is therefore

$$y = y_h + y_p$$
$$= c_1 + c_2 \cos x + c_3 \sin x + \ln|\sec x + \tan x| - x\cos x + (\sin x)\ln|\cos x|$$

Chapter 7

APPLICATIONS OF SECOND-ORDER LINEAR DIFFERENTIAL EQUATIONS

IN THIS CHAPTER:

✔ *Spring Problems*
✔ *Electrical Circuit Problems*
✔ *Buoyancy Problems*
✔ *Classifying Solutions*
✔ *Solved Problems*

Spring Problems

The simple spring system shown in Figure 7-1 consists of a mass m attached to the lower end of a spring that is itself suspended vertically from a mounting. The system is in its *equilibrium position* when it is at rest. The mass is set in motion by one or more of the following means: displacing the mass from its equilibrium position, providing it with an initial velocity, or subjecting it to an external force $F(t)$.

47

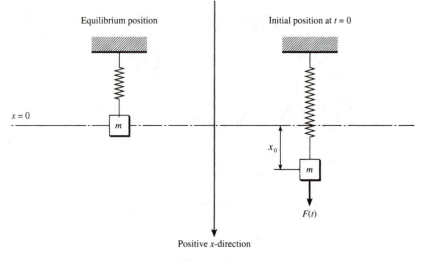

Equilibrium position

Initial position at $t = 0$

$x = 0$

m

x_0

m

$F(t)$

Positive x-direction

Figure 7-1

Hooke's law: *The restoring force F of a spring is equal and opposite to the forces applied to the spring and is proportional to the extension (contraction) l of the spring as a result of the applied force; that is, $F = -kl$, where k denotes the constant of proportionality, generally called the spring constant.*

Example 7.1. A steel ball weighing 128 lb is suspended from a spring, whereupon the spring is stretched 2 ft from its natural length. The applied force responsible for the 2-ft displacement is the weight of the ball, 128 lb. Thus, $F = -128$ lb. Hooke's law then gives $-128 = -k(2)$, or $k = 64$ lb/ft.

For convenience, we choose the downward direction as the positive direction and take the origin to be the center of gravity of the mass in the equilibrium position. We assume that the mass of the spring is negligible and can be neglected and that air resistance, when present, is proportional to the velocity of the mass. Thus, at any time t, there are three forces acting on the system: (1) $F(t)$, measured in the positive direction; (2) a restoring force given by Hooke's law as $F_s = -kx$, $k > 0$; and (3) a force due to air resistance given by $F_a = -a\dot{x}$, $a > 0$, where a is the constant of proportionality. Note that the restoring force F_s always acts in a direction that will tend to return the system to the equilibrium position: if the mass is below the equilibrium position, then x is positive and $-kx$ is negative; whereas if the mass is above the equilibrium position, then x is negative and $-kx$ is positive. Also note that because $a > 0$ the force F_a due to air resistance acts in the opposite direction of the velocity and thus tends to retard, or damp, the motion of the mass.

It now follows from Newton's second law (see Chapter Three) that $m\ddot{x} = -kx - a\dot{x} + F(t)$, or

$$\ddot{x} + \frac{a}{m}\dot{x} + \frac{k}{m}x = \frac{F(t)}{m} \qquad (7.1)$$

If the system starts at $t = 0$ with an initial velocity v_0 and from an initial position x_0, we also have the initial conditions

$$x(0) = x_0 \qquad \dot{x}(0) = v_0 \qquad (7.2)$$

The force of gravity does not explicitly appear in 7.1, but it is present nonetheless. We automatically compensated for this force by measuring distance from the equilibrium position of the spring. If one wishes to exhibit gravity explicitly, then distance must be measured from the bottom end of the *natural length* of the spring. That is, the motion of a vibrating spring can be given by

$$\ddot{x} + \frac{a}{m}\dot{x} + \frac{k}{m}x = g + \frac{F(t)}{m}$$

if the origin, $x = 0$, is the terminal point of the unstretched spring before the mass m is attached.

Electrical Circuit Problems

The simple electrical circuit shown in Figure 7-2 consists of a resistor R in ohms; a capacitor C in farads; an inductor L in henries; and an electromotive force (emf) $E(t)$ in volts, usually a battery or a generator, all connected in series. The current I flowing through the circuit is measured in amperes and the charge q on the capacitor is measured in coulombs.

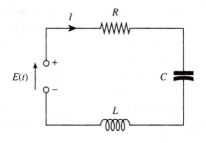

Figure 7-2

Kirchhoff's loop law: *The algebraic sum of the voltage drops in a simple closed electric circuit is zero.*

It is known that the voltage drops across a resistor, a capacitor, and an inductor are respectively RI, $(1/C)q$, and $L(dI/dt)$ where q is the charge on the capacitor. The voltage drop across an emf is $-E(t)$. Thus, from Kirchhoff's loop law, we have

$$RI + L\frac{dI}{dt} + \frac{1}{C}q - E(t) = 0 \tag{7.3}$$

The relationship between q and I is

$$I = \frac{dq}{dt} \qquad \frac{dI}{dt} = \frac{d^2q}{dt^2} \tag{7.4}$$

Substituting these values into Equation 7.3, we obtain

$$\frac{d^2q}{dt^2} + \frac{R}{L}\frac{dq}{dt} + \frac{1}{LC}q = \frac{1}{L}E(t) \tag{7.5}$$

The initial conditions for q are

$$q(0) = q_0 \qquad \left.\frac{dq}{dt}\right|_{t=0} = I(0) = I_0 \tag{7.6}$$

To obtain a differential equation for the current, we differentiate Equation 7.3 with respect to t and then substitute Equation 7.4 directly into the resulting equation to obtain

$$\frac{d^2I}{dt^2} + \frac{R}{L}\frac{dI}{dt} + \frac{1}{LC}I = \frac{1}{L}\frac{dE(t)}{dt} \tag{7.7}$$

The first initial condition is $I(0) = I_0$. The second initial condition is obtained from Equation 7.3 by solving for dI/dt and then setting $t = 0$. Thus,

$$\left.\frac{dI}{dt}\right|_{t=0} = \frac{1}{L}E(0) - \frac{R}{L}I_0 - \frac{1}{LC}q_0 \qquad (7.8)$$

An expression for the current can be gotten either by solving Equation 7.7 directly or by solving Equation 7.5 for the charge and then differentiating that expression.

Buoyancy Problems

Consider a body of mass m submerged either partially or totally in a liquid of weight density ρ. Such a body experiences two forces, a downward force due to gravity and a counter force governed by:

Archimedes' principle: *A body in liquid experiences a buoyant upward force equal to the weight of the liquid displaced by that body.*

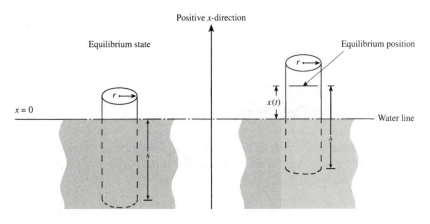

Figure 7-3

Equilibrium occurs when the buoyant force of the displaced liquid equals the force of gravity on the body. Figure 7-3 depicts the situation for a cylinder of radius r and height H where h units of cylinder height are submerged at equilibrium. At equilibrium, the volume of water displaced by the cylinder is $\pi r^2 h$, which provides a buoyant force of $\pi r^2 h \rho$ that must equal the weight of the cylinder mg. Thus,

$$\pi r^2 h \rho = mg \tag{7.9}$$

Motion will occur when the cylinder is displaced from its equilibrium position. We arbitrarily take the upward direction to be the positive x-direction. If the cylinder is raised out of the water by $x(t)$ units, as shown in Figure 7-3, then it is no longer in equilibrium. The downward or negative force on such a body remains mg but the buoyant or positive force is reduced to $\pi r^2 [h - x(t)] \rho$. It now follows from Newton's second law that

$$m\ddot{x} = \pi r^2 [h - x(t)] \rho - mg$$

Substituting 7.9 into this last equation, we can simplify it to

$$m\ddot{x} = -\pi r^2 x(t) \rho$$

or

$$\ddot{x} + \frac{\pi r^2 \rho}{m} x = 0 \tag{7.10}$$

Classifying Solutions

Vibrating springs, simple electrical circuits, and floating bodies are all governed by second-order linear differential equations with constant coefficients of the form

$$\ddot{x} + a_1 \dot{x} + a_0 x = f(t) \tag{7.11}$$

For vibrating spring problems defined by Equation 7.1, $a_1 = a/m$, $a_0 = k/m$, and $f(t) = F(t)/m$. For buoyancy problems defined by Equation 7.10, $a_1 = 0$, $a_0 = \pi r^2 \rho/m$, and $f(t) \equiv 0$. For electrical circuit problems, the independent variable x is replaced either by q in Equation 7.5 or I in Equation 7.7.

The motion or current in all of these systems is classified as *free* and *undamped* when $f(t) \equiv 0$ and $a_1 = 0$. It is classified as *free* and *damped* when $f(t)$ is identically zero but a_1 is not zero. For damped motion, there are three separate cases to consider, according as the roots of the associated characteristic equation (see Chapter Five) are (1) real and distinct, (2) equal, or (3) complex conjugate. These cases are respectively classified as (1) *overdamped*, (2) *critically damped*, and (3) *os-*  *cillatory damped* (or, in electrical problems, *underdamped*). If $f(t)$ is not identically zero, the motion or current is classified as *forced*.

A motion or current is *transient* if it "dies out" (that is, goes to zero) as $t \rightarrow \infty$. A *steady-state* motion or current is one that is not transient and does not become unbounded. Free damped systems always yield transient motions, while forced damped systems (assuming the external force to be sinusoidal) yield both transient and steady-state motions.

Free undamped motion defined by Equation 7.11 with $a_1 = 0$ and $f(t) \equiv 0$ always has solutions of the form

$$x(t) = c_1 \cos \omega t + c_2 \sin \omega t \qquad (7.12)$$

which defines *simple harmonic motion*. Here c_1, c_2, and ω are constants with ω often referred to as *circular frequency*. The *natural frequency f* is

$$f = \frac{\omega}{2\pi}$$

and it represents the number of complete oscillations per time unit undertaken by the solution. The *period* of the system of the time required to complete one oscillation is

$$T = \frac{1}{f}$$

Equation 7.12 has the alternate form

$$x(t) = (-1)^k A \cos(\omega t - \phi) \qquad (7.13)$$

where the *amplitude* $A = \sqrt{c_1^2 + c_2^2}$, the *phase angle* $\phi = \arctan(c_2/c_1)$, and k is zero when c_1 is positive and unity when c_1 is negative.

Solved Problems

Solved Problem 7.1 A 10-kg mass is attached to a spring, stretching it 0.7 m from its natural length. The mass is started in motion from the equilibrium position with an initial velocity of 1 m/sec in the upward direction. Find the subsequent motion, if the force due to air resistance is $-90\dot{x}$ N.

Taking $g = 9.8\text{m/sec}^2$, we have $w = mg = 98$ N and $k = w/l = 140$ N/m. Furthermore, $a = 90$ and $F(t) \equiv 0$ (there is no external force). Equation 7.1 becomes

$$\ddot{x} + 9\dot{x} + 14x = 0 \tag{7.14}$$

The roots of the associated characteristic equation are $\lambda_1 = -2$ and $\lambda_2 = -7$, which are real and distinct; hence this problem is an example of overdamped motion. The solution of 7.14 is

$$x = c_1 e^{-2t} + c_2 e^{-7t}$$

The initial conditions are $x(0) = 0$ (the mass starts at the equilibrium position) and $\dot{x}(0) = -1$ (the initial velocity is in the negative direction). Applying these conditions, we find that $c_1 = -c_2 = -\frac{1}{5}$, so that $x = \frac{1}{5}(e^{-7t} - e^{-2t})$. Note that $x \to 0$ as $t \to \infty$; thus, the motion is transient.

Solved Problem 7.2 Determine whether a cylinder of radius 4 in, height 10 in, and weight 15 lb can float in a deep pool of water of weight density 62.5 lb/ft³.

Let h denote the length (in feet) of the submerged portion of the cylinder at equilibrium. With $r = \frac{1}{3}$ ft, it follows from Equation 7.9 that

$$h = \frac{mg}{\pi r^2 \rho} = \frac{15}{\pi(\frac{1}{3})^2 62.5} \approx 0.688\text{ft} = 8.25\text{in}$$

Thus, the cylinder will float with $10 - 8.25 = 1.75$ in of length above the water at equilibrium.

Chapter 8

LAPLACE TRANSFORMS AND INVERSE LAPLACE TRANSFORMS

IN THIS CHAPTER:

Definition of the Laplace Transform

Let $f(x)$ be defined for $0 \le x < \infty$ and let s denote an arbitrary real variable. The *Laplace transform of* $f(x)$, designated by either $\mathcal{L}\{f(x)\}$ or $F(s)$, is

$$\mathcal{L}\{f(x)\} = F(s) = \int_0^\infty e^{-sx} f(x)dx \tag{8.1}$$

for all values of s for which the improper integral converges. Convergence occurs when the limit

$$\lim_{R \to \infty} \int_0^R e^{-sx} f(x)dx \tag{8.2}$$

exists. If this limit does not exist, the improper integral diverges and $f(x)$ has no Laplace transform. When evaluating the integral in Equation 8.1, the variable s is treated as a constant because the integration is with respect to x.

The Laplace transforms for a number of elementary functions are given in Appendix A.

Properties of Laplace Transforms

Property 8.1 (Linearity). If $\mathcal{L}\{f(x)\} = F(s)$ and $\mathcal{L}\{g(x)\} = G(s)$, then for any two constants c_1 and c_2

$$\begin{aligned} \mathcal{L}\{c_1 f(x) + c_2 g(x)\} &= c_1 \mathcal{L}\{f(x)\} + c_2 \mathcal{L}\{g(x)\} \\ &= c_1 F(s) + c_2 G(s) \end{aligned} \tag{8.3}$$

Property 8.2. If $\mathcal{L}\{f(x)\} = F(s)$, then for any constant a

$$\mathcal{L}\{e^{ax}f(x)\} = F(s - a) \tag{8.4}$$

Property 8.3. If $\mathcal{L}\{f(x)\} = F(s)$, then for any positive integer n

$$\mathcal{L}\{x^n f(x)\} = (-1)^n \frac{d^n}{ds^n}[F(s)] \tag{8.5}$$

Property 8.4. If $\mathcal{L}\{f(x)\} = F(s)$ and if $\lim\limits_{\substack{x\to 0 \\ x>0}} \dfrac{f(x)}{x}$ exists, then

$$\mathcal{L}\left\{\frac{1}{x}f(x)\right\} = \int_s^\infty F(t)dt \tag{8.6}$$

Property 8.5. If $\mathcal{L}\{f(x)\} = F(s)$, then

$$\mathcal{L}\left\{\int_0^x f(t)dt\right\} = \frac{1}{s}F(s) \tag{8.7}$$

Property 8.6. If $f(x)$ is periodic with period ω, that is, $f(x+\omega)=f(x)$, then

$$\mathcal{L}\{f(x)\} = \frac{\displaystyle\int_0^\omega e^{-sx}f(x)dx}{1-e^{-\omega s}} \tag{8.8}$$

Functions of Other Independent Variables

For consistency only, the definition of the Laplace transform and its properties, Equations 8.1 through 8.8, are presented for functions of x. They are equally applicable for functions of any independent variable and are generated by replacing the variable x in the above equations by any variable of interest. In particular, the counter part of Equation 8.1 for the Laplace transform of a function of t is

$$\mathcal{L}\{f(t)\} = F(s) = \int_0^\infty e^{-st}f(t)dt$$

Definition of the Inverse Laplace Transform

An inverse *Laplace transform* of $F(s)$ designated by $\mathcal{L}^{-1}\{F(s)\}$, is another function $f(x)$ having the property that $\mathcal{L}\{f(x)\} = F(s)$.

The simplest technique for identifying inverse Laplace transforms is to recognize them, either from memory or from a table such as in the Appendix. If $F(s)$ is not in a recognizable form, then occasionally it can be transformed into such a form by algebraic manipulation. Observe from the Appendix that almost all Laplace transforms are quotients. The recommended procedure is to first convert the denominator to a form that appears in the Appendix and then the numerator.

Manipulating Denominators

The method of *completing the square* converts a quadratic polynomial into the sum of squares, a form that appears in many of the denominators in the Appendix. In particular, for the quadratic

$$
\begin{aligned}
as^2 + bs + c &= a\left(s^2 + \frac{b}{a}s\right) + c \\
&= a\left[s^2 + \frac{b}{a}s + \left(\frac{b}{2a}\right)^2\right] + \left[c - \frac{b^2}{4a}\right] \\
&= a\left(s + \frac{b}{2a}\right)^2 + \left(c - \frac{b^2}{4a}\right) \\
&= a(s+k)^2 + h^2
\end{aligned}
$$

where $k = b/2a$ and $h = \sqrt{c - (b^2/4a)}$.

The method of *partial fractions* transforms a function of the form $a(s)/b(s)$, where both $a(s)$ and $b(s)$ are polynomials in s, into the sum of other fractions such that the denominator of each new fraction is either a first-degree or a quadratic polynomial raised to some power. The method requires only that the degree of $a(s)$ be less than the degree of $b(s)$ (if this is not the case, first perform long division, and consider the remainder term) and $b(s)$ be factored into the product of distinct linear and quadratic polynomials raised to various powers.

The method is carried out as follows. To each factor of $b(s)$ of the form $(s - a)^m$, assign a sum of m fractions, of the form

$$
\frac{A_1}{s-a} + \frac{A_2}{(s-a)^2} + \cdots + \frac{A_m}{(s-a)^m}
$$

To each factor of $b(s)$ of the form $(s^2 + bs + c)^p$, assign a sum of p fractions, of the form

$$
\frac{B_1 s + C_1}{s^2 + bs + c} + \frac{B_2 s + C_2}{(s^2 + bs + c)^2} + \cdots + \frac{B_p s + C_p}{(s^2 + bs + c)^p}
$$

Here A_i, B_j, and C_k $(i = 1,2,..., m; j, k = 1,2,..., p)$ are constants which still must be determined.

Set the original fraction $a(s)/b(s)$ equal to the sum of the new fractions just constructed. Clear the resulting equation of fractions and then equate coefficients of like powers of s, thereby obtaining a set of simultaneous linear equations in the unknown constants A_i, B_j, and C_k. Finally, solve these equations for A_i, B_j, and C_k.

Manipulating Numerators

A factor $s - a$ in the numerator may be written in terms of the factor $s - b$, where both a and b are constants, through the identity $s - a = (s - b) + (b - a)$. The multiplicative constant a in the numerator may be written explicitly in terms of the multiplicative constant b through the identity

$$a = \frac{a}{b}(b)$$

Both identities generate recognizable inverse Laplace transforms when they are combined with:

Property 8.7 (Linearity). If the inverse Laplace transforms of two functions $F(s)$ and $G(s)$ exist, then for any constants c_1 and c_2,

$$\mathcal{L}^{-1}\{c_1 F(s) + c_2 G(s)\} = c_1 \mathcal{L}^{-1}\{F(s)\} + c_2 \mathcal{L}^{-1}\{G(s)\}$$

Convolutions

The *convolution* of two functions $f(x)$ and $g(x)$ is

$$f(x) * g(x) = \int_0^x f(t)g(x-t)dt$$

(8.9)

Theorem 8.1. $f(x) * g(x) = g(x) * f(x)$.

Theorem 8.2. (*Convolution Theorem*). If $\mathcal{L}\{f(x)\} = F(s)$ and $\mathcal{L}\{g(x)\} = G(s)$, then $\mathcal{L}\{f(x) * g(x)\} = \mathcal{L}\{f(x)\} \mathcal{L}\{g(x)\} = F(s)G(s)$

You Need to Know ✔

The inverse Laplace transform of a product is computed using a convolution.

$$\mathcal{L}^{-1}\{F(s)G(s)\} = f(x) * g(x) = g(x) * f(x) \qquad (8.10)$$

If one of the two convolutions in Equation 8.10 is simpler to calculate, then that convolution is chosen when determining the inverse Laplace transform of a product.

Unit Step Function

The *unit step function* $u(x)$ is defined as

$$u(x) = \begin{cases} 0 & x < 0 \\ 1 & x \geq 0 \end{cases}$$

As an immediate consequence of the definition, we have for any number c,

$$u(x - c) = \begin{cases} 0 & x < c \\ 1 & x \geq c \end{cases}$$

The graph of $u(x - c)$ is given in Figure 8-1.

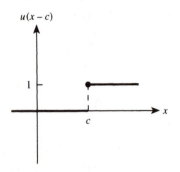

Figure 8-1

Theorem 8.3. $\mathcal{L}\{u(x-c)\} = \frac{1}{s}e^{-cs}.$

Translations

Given a function $f(x)$ defined for $x \ge 0$, the function

$$u(x-c)f(x-c) = \begin{cases} 0 & x < c \\ f(x-c) & x \ge c \end{cases}$$

represents a shift, or translation, of the function $f(x)$ by c units in the positive x-direction. For example, if $f(x)$ is given graphically by Figure 8-2, then $u(x-c)f(x-c)$ is given graphically by Figure 8-3.

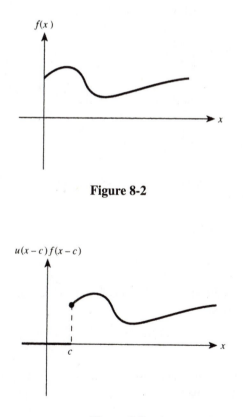

Figure 8-2

Figure 8-3

Theorem 8.4. If $F(s) = \mathcal{L}\{f(x)\}$, then

$$\mathcal{L}\{u(x-c)f(x-c)\} = e^{-cs}F(s)$$

Conversely,

$$\mathcal{L}^{-1}\{e^{-cs}F(s)\} = u(x-c)f(x-c) = \begin{cases} 0 & x < c \\ f(x-c) & x \geq c \end{cases}$$

Solved Problems

Solved Problem 8.1 Find $\mathcal{L}\{e^{ax}\}$.

Using Equation 8.1, we obtain

$$F(s) = \int_0^\infty e^{-sx}e^{ax}dx = \lim_{R\to\infty}\int_0^R e^{(a-s)x}dx$$

$$= \lim_{R\to\infty}\left[\frac{e^{(a-s)x}}{a-s}\right]_{x=0}^{x=R} = \lim_{R\to\infty}\left[\frac{e^{(a-s)R}-1}{a-s}\right]$$

$$= \frac{1}{s-a}\,(\text{for } s > a)$$

Note that when $s \leq a$, the improper integral diverges. (See also entry 7 in the Appendix.)

Solved Problem 8.2 Find $\mathcal{L}\{xe^{4x}\}$.

This problem can be done three ways.

(a) Using entry 14 of the Appendix with $n = 2$ and $a = 4$, we have directly that

$$\mathcal{L}\{xe^{4x}\} = \frac{1}{(s-4)^2}$$

(b) Set $f(x) = x$. Using Property 8.2 with $a = 4$ and entry 2 of the Appendix, we have

$$F(s) = \mathcal{L}\{f(x)\} = \mathcal{L}\{x\} = \frac{1}{s^2}$$

and

$$\mathcal{L}\{e^{4x}x\} = F(s-4) = \frac{1}{(s-4)^2}$$

(c) Set $f(x) = e^{4x}$. Using Property 8.3 with $n = 1$ and the results of Problem 8.1, or alternatively, entry 7 of the Appendix with $a = 4$ we find that

$$F(s) = \mathcal{L}\{f(x)\} = \mathcal{L}\{e^{4x}\} = \frac{1}{s-4}$$

and

$$\mathcal{L}\{xe^{4x}\} = -F'(s) = -\frac{d}{ds}\left(\frac{1}{s-4}\right) = \frac{1}{(s-4)^2}$$

Solved Problem 8.3 Use partial fractions to decompose $\dfrac{s+3}{(s-2)(s+1)}$.

To the linear factors $s - 2$ and $s + 1$, we associate respectively the fractions $A/(s - 2)$ and $B/(s + 1)$. We set

$$\frac{s+3}{(s-2)(s+1)} \equiv \frac{A}{s-2} + \frac{B}{s+1}$$

and, upon clearing fractions, obtain

$$s+3 \equiv A(s+1) + B(s-2) \tag{8.11}$$

To find A and B, we substitute $s = -1$ and $s = 2$ into 8.11, we immediately obtain $A = 5/3$ and $B = -2/3$. Thus,

$$\frac{s+3}{(s-2)(s+1)} \equiv \frac{5/3}{s-2} - \frac{2/3}{s+1}$$

Solved Problem 8.4 Find $\mathcal{L}^{-1}\left\{\dfrac{s+3}{(s-2)(s+1)}\right\}$.

No function of this form appears in the Appendix. Using the results of Problem 8.3 and Property 8.7, we obtain

$$\mathcal{L}^{-1}\left\{\frac{s+3}{(s-2)(s+1)}\right\} = \frac{5}{3}\mathcal{L}^{-1}\left\{\frac{1}{s-2}\right\} - \frac{2}{3}\mathcal{L}^{-1}\left\{\frac{1}{s+1}\right\}$$

$$= \frac{5}{3}e^{2x} - \frac{2}{3}e^{-x}$$

Chapter 9
SOLUTIONS BY LAPLACE TRANSFORMS

IN THIS CHAPTER:

✔ *Laplace Transforms of Derivatives*
✔ *Solutions of Linear Differential Equations with Constant Coefficients*
✔ *Solutions of Linear Systems*
✔ *Solved Problems*

Laplace Transforms of Derivatives

Denote $\mathcal{L}\{y(x)\}$ by $Y(s)$. Then under very broad conditions, the Laplace transform of the nth-derivative ($n = 1,2,3,...$) of $y(x)$ is

$$\mathcal{L}\left\{\frac{d^n y}{dx^n}\right\} = s^n Y(s) - s^{n-1}y(0) - s^{n-2}y'(0) - \\ \cdots - s y^{(n-2)}(0) - y^{(n-1)}(0) \tag{9.1}$$

If the initial conditions on $y(x)$ at $x = 0$ are given by

$$y(0) = c_0, y'(0) = c_1, ..., y^{(n-1)}(0) = c_{n-1} \tag{9.2}$$

then (9.1) can be rewritten as

$$\mathcal{L}\left\{\frac{d^n y}{dx^n}\right\} = s^n Y(s) - c_0 s^{n-1} - c_1 s^{n-2} - \cdots - c_{n-2}s - c_{n-1} \quad (9.3)$$

For the special cases of $n = 1$ and $n = 2$, Equation 9.3 simplifies to

$$\mathcal{L}\{y'(x)\} = sY(s) - c_0 \quad (9.4)$$

$$\mathcal{L}\{y''(x)\} = s^2 Y(s) - c_0 s - c_1 \quad (9.5)$$

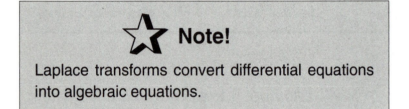

☆ Note!

Laplace transforms convert differential equations into algebraic equations.

Solutions of Linear Differential Equations with Constant Coefficients

Laplace transforms are used to solve initial-value problems given by the nth-order linear differential equation with constant coefficients

$$b_n \frac{d^n y}{dx^n} + b_{n-1} \frac{d^{n-1} y}{dx^{n-1}} + \cdots + b_1 \frac{dy}{dx} + b_0 y = g(x) \quad (9.6)$$

together with the initial conditions specified in Equation 9.2. First, take the Laplace transform of both sides of Equation 9.6, thereby obtaining an algebraic equation for $Y(s)$. Then solve for $Y(s)$ *algebraically*, and finally take inverse Laplace transforms to obtain $y(x) = \mathcal{L}^{-1}\{Y(s)\}$.

Unlike previous methods, where first the differential equation is solved and then the initial conditions are applied to evaluate the arbitrary constants, the Laplace transform method solves the entire initial-value problem in one step. There are two exceptions: when no initial conditions are specified and when the initial conditions are not at $x = 0$. In these situations, c_0 through c_{n-1} in Equations 9.2 and 9.3 remain arbitrary

and the solution to differential equation 9.6 is found in terms of these constants. They are then evaluated separately when appropriate subsidiary conditions are provided.

Solutions of Linear Systems

Laplace transforms are useful for solving systems of linear differential equations; that is, sets of two or more differential equations with an equal number of unknown functions. If all of the coefficients are constants, then the method of solution is a straightforward generalization of the one described above. Laplace transforms are taken of each differential equation in the system; the transforms of the unknown functions are determined algebraically from the resulting set of simultaneous equations; inverse transforms for the unknown functions are calculated with the help of the Appendix.

Solved Problems

Solved Problem 9.1 Solve $y' - 5y = e^{5x}$; $y(0) = 0$.

Taking the Laplace transform of both sides of this differential equation and using Property 8.1, we find that $\mathcal{L}\{y'\} - 5\mathcal{L}\{y\} = \mathcal{L}\{e^{5x}\}$. Then, using the Appendix and Equation 9.4 with $c_0 = 0$, we obtain

$$[sY(s) - 0] - 5Y(s) = \frac{1}{s-5} \text{ from which } Y(s) = \frac{1}{(s-5)^2}$$

Finally, taking the inverse Laplace transform of $Y(s)$, we obtain

$$y(x) = \mathcal{L}^{-1}\{Y(s)\} = \mathcal{L}^{-1}\left\{\frac{1}{(s-5)^2}\right\} = xe^{5x}$$

(see Appendix, entry 14).

Solved Problem 9.2 Solve the system

$$y'' + z + y = 0$$
$$z' + y' = 0;$$
$$y(0) = 0, \quad y'(0) = 0, \quad z(0) = 1$$

Denote $\mathcal{L}\{y(x)\}$ and $\mathcal{L}\{z(x)\}$ by $Y(s)$ and $Z(s)$ respectively. Then, taking the Laplace transforms of both differential equations, we obtain

$$[s^2 Y(s) - (0)s - (0)] + Z(s) + Y(s) = 0$$
$$[sZ(s) - 1] + [sY(s) - 0] = 0$$

or

$$(s^2 + 1)Y(s) + Z(s) = 0$$
$$Y(s) + Z(s) = \frac{1}{s}$$

Solving this last system for $Y(s)$ and $Z(s)$, we find that

$$Y(s) = -\frac{1}{s^3} \qquad Z(s) = \frac{1}{s} + \frac{1}{s^3}$$

Thus, taking inverse transforms, we conclude that

$$y(x) = -\frac{1}{2}x^2 \qquad z(x) = 1 + \frac{1}{2}x^2$$

Chapter 10

MATRICES AND THE MATRIX EXPONENTIAL

IN THIS CHAPTER:

- ✔ *Matrices and Vectors*
- ✔ *Matrix Addition*
- ✔ *Scalar and Matrix Multiplication*
- ✔ *Powers of a Square Matrix*
- ✔ *Differentiation and Integration of Matrices*
- ✔ *The Characteristic Equation of a Matrix*
- ✔ *Definition of the Matrix Exponential e^{At}*
- ✔ *Computation of the Matrix Exponential e^{At}*
- ✔ *Solved Problems*

Matrices and Vectors

A *matrix* (designated by an uppercase boldface letter) is a rectangular array of elements arranged in horizontal rows and vertical columns. In this book, the elements of matrices will always be numbers or functions of the variable t. If all the elements are numbers, then the matrix is called a *constant matrix*.

Example 10.1.

$$\begin{bmatrix} 1 & 2 \\ 3 & 4 \end{bmatrix}, \begin{bmatrix} 1 & e^t & 2 \\ t & -1 & 1 \end{bmatrix}, \text{ and } \begin{bmatrix} 1 & t^2 & \cos t \end{bmatrix}$$

are all matrices. In particular, the first matrix is a constant matrix, whereas the last two are not.

A general matrix $\mathbf{A}$ having p rows and n columns is given by

$$\mathbf{A} = \begin{bmatrix} a_{ij} \end{bmatrix} = \begin{bmatrix} a_{11} & a_{12} & \cdots & a_{1n} \\ a_{21} & a_{22} & \cdots & a_{2n} \\ \vdots & \vdots & & \vdots \\ a_{p1} & a_{p2} & \cdots & a_{pn} \end{bmatrix}$$

where a_{ij} represents that element appearing in the ith row and jth column.

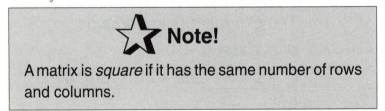

Note!

A matrix is *square* if it has the same number of rows and columns.

A *vector* (designated by a lowercase boldface letter) is a matrix having only one column or one row. (The third matrix given in Example 10.1 is a vector.)

Matrix Addition

The *sum* $\mathbf{A} + \mathbf{B}$ of two matrices $\mathbf{A} = [a_{ij}]$ and $\mathbf{B} = [b_{ij}]$ having the same number of rows and the same number of columns is the matrix obtained by adding the corresponding elements of $\mathbf{A}$ and $\mathbf{B}$. That is,

$$\mathbf{A} + \mathbf{B} = [a_{ij}] + [b_{ij}] = [a_{ij} + b_{ij}]$$

Matrix addition is both associative and commutative. Thus, $\mathbf{A} + (\mathbf{B} + \mathbf{C}) = (\mathbf{A} + \mathbf{B}) + \mathbf{C}$ and $\mathbf{A} + \mathbf{B} = \mathbf{B} + \mathbf{A}$.

Scalar and Matrix Multiplication

If λ is either a number or a function of t, then $\lambda\mathbf{A}$ (or, equivalently, $\mathbf{A}\lambda$) is defined to be the matrix obtained by multiplying every element of $\mathbf{A}$ by λ. That is,

$$\lambda\mathbf{A} = \lambda[a_{ij}] = [\lambda a_{ij}]$$

Let $\mathbf{A} = [a_{ij}]$ and $\mathbf{B} = [b_{ij}]$ be two matrices such that $\mathbf{A}$ has r rows and n columns and $\mathbf{B}$ has n rows and p columns. Then the *product* $\mathbf{AB}$ is defined to be the matrix $\mathbf{C} = [c_{ij}]$ given by

$$c_{ij} = \sum_{k=1}^{n} a_{ik}b_{kj} \qquad (i = 1, 2, \ldots, r; j = 1, 2, \ldots, p)$$

The element c_{ij} is obtained by multiplying the elements of the ith row of $\mathbf{A}$ with the corresponding elements of the jth column of $\mathbf{B}$ and summing the results.

Matrix multiplication is associative and distributes over addition; in general, however, it is *not* commutative. Thus,

$\mathbf{A}(\mathbf{BC}) = (\mathbf{AB})\mathbf{C}$, $\mathbf{A}(\mathbf{B} + \mathbf{C}) = \mathbf{AB} + \mathbf{AC}$, and $(\mathbf{B} + \mathbf{C})\mathbf{A} = \mathbf{BA} + \mathbf{CA}$.

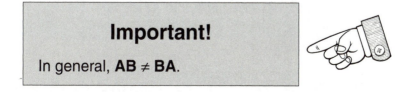

Important!

In general, $\mathbf{AB} \neq \mathbf{BA}$.

Powers of a Square Matrix

If n is a positive integer and $\mathbf{A}$ is a square matrix, then

$$\mathbf{A}^n = \underbrace{\mathbf{A}\mathbf{A}\cdots\mathbf{A}}_{n \ \ times}$$

In particular, $\mathbf{A}^2 = \mathbf{A}\mathbf{A}$ and $\mathbf{A}^3 = \mathbf{A}\mathbf{A}\mathbf{A}$. By definition, $\mathbf{A}^0 = \mathbf{I}$, where

$$\mathbf{I} = \begin{bmatrix} 1 & 0 & 0 & \cdots & 0 & 0 \\ 0 & 1 & 0 & \cdots & 0 & 0 \\ 0 & 0 & 1 & \cdots & 0 & 0 \\ \vdots & & & \ddots & & \vdots \\ 0 & 0 & 0 & \cdots & 1 & 0 \\ 0 & 0 & 0 & \cdots & 0 & 1 \end{bmatrix}$$

is called an *identity matrix*. For any square matrix $\mathbf{A}$ and identity matrix $\mathbf{I}$ of the same size

$$\mathbf{A}\mathbf{I} = \mathbf{I}\mathbf{A} = \mathbf{A}$$

Differentiation and Integration of Matrices

The *derivative* of $\mathbf{A} = [a_{ij}]$ is the matrix obtained by differentiating each element of $\mathbf{A}$; that is,

$$\frac{d\mathbf{A}}{dt} = \left[\frac{da_{ij}}{dt} \right]$$

Similarly, the *integral* of $\mathbf{A}$, either definite or indefinite, is obtained by integrating each element of $\mathbf{A}$. Thus,

$$\int_a^b \mathbf{A}dt = \left[\int_a^b a_{ij}dt \right] \quad \text{and} \quad \int \mathbf{A}dt = \left[\int a_{ij}dt \right]$$

The Characteristic Equation of a Matrix

The *characteristic equation* of a square matrix $\mathbf{A}$ is the polynomial equation in λ given by

$$\det(\mathbf{A} - \lambda\mathbf{I}) = 0 \qquad (10.1)$$

where det( ) stands for "the determinant of." Those values of λ which satisfy 10.1, that is, the roots of 10.1, are the *eigenvalues* of $\mathbf{A}$, a k-fold root being called an *eigenvalue of multiplicity k*.

Theorem 10.1. (*Cayley-Hamilton theorem*). Any square matrix satisfies its own characteristic equation. That is, if

$$\det(\mathbf{A} - \lambda\mathbf{I}) = b_n\lambda^n + b_{n-1}\lambda^{n-1} + \cdots + b_2\lambda^2 + b_1\lambda + b_0$$

then $\qquad b_n\mathbf{A}^n + b_{n-1}\mathbf{A}^{n-1} + \cdots + b_2\mathbf{A}^2 + b_1\mathbf{A} + b_0\mathbf{I} = \mathbf{0}.$

Definition of the Matrix Exponential $e^{\mathbf{A}t}$

For a square matrix $\mathbf{A}$,

$$e^{\mathbf{A}t} \equiv \mathbf{I} + \frac{1}{1!}\mathbf{A}t + \frac{1}{2!}\mathbf{A}^2t^2 + \cdots = \sum_{n=0}^{\infty} \frac{1}{n!}\mathbf{A}^nt^n \qquad (10.2)$$

The infinite series 10.2 converges for every $\mathbf{A}$ and t, so that $e^{\mathbf{A}t}$ is defined for all square matrices.

Computation of the Matrix Exponential $e^{\mathbf{A}t}$

For actually computing the elements of $e^{\mathbf{A}t}$, 10.2 is not generally useful. However, it follows (with some effort) from Theorem 10.1, applied to the matrix $\mathbf{A}t$, that the infinite series can be reduced to a polynomial in t. Thus:

Theorem 10.2. If $\mathbf{A}$ is a matrix having n rows and n columns, then

$$e^{\mathbf{A}t} = \alpha_{n-1}\mathbf{A}^{n-1}t^{n-1} + \alpha_{n-2}\mathbf{A}^{n-2}t^{n-2} + \cdots + \alpha_1\mathbf{A}t + \alpha_0\mathbf{I} \qquad (10.3)$$

where α_0, α_1,...,α_{n-1} are functions of t which must be determined for each $\mathbf{A}$.

Example 10.2. When $\mathbf{A}$ has two rows and two columns, then $n = 2$ and

$$e^{\mathbf{A}t} = \alpha_1 \mathbf{A}t + \alpha_0 \mathbf{I} \tag{10.4}$$

When $\mathbf{A}$ has three rows and three columns, then $n = 3$ and

$$e^{\mathbf{A}t} = \alpha_2 \mathbf{A}^2 t^2 + \alpha_1 \mathbf{A}t + \alpha_0 \mathbf{I} \tag{10.5}$$

Theorem 10.3. Let $\mathbf{A}$ be as in Theorem 10.2, and define

$$r(\lambda) \equiv \alpha_{n-1}\lambda^{n-1} + \alpha_{n-2}\lambda^{n-2} + \cdots + \alpha_2\lambda^2 + \alpha_1\lambda + \alpha_0 \tag{10.6}$$

Then if λ_i is an eigenvalue of $\mathbf{A}t$,

$$e^{\lambda_i} = r(\lambda_i) \tag{10.7}$$

Furthermore, if λ_i is an eigenvalue of multiplicity k, $k > 1$, then the following equations are also valid:

$$
\begin{aligned}
e^{\lambda_i} &= \frac{d}{d\lambda} r(\lambda)\Big|_{\lambda=\lambda_i} \\
e^{\lambda_i} &= \frac{d^2}{d\lambda^2} r(\lambda)\Big|_{\lambda=\lambda_i} \\
&\cdots\cdots\cdots\cdots\cdots \\
e^{\lambda_i} &= \frac{d^{k-1}}{d\lambda^{k-1}} r(\lambda)\Big|_{\lambda=\lambda_i}
\end{aligned}
\tag{10.8}
$$

Note that Theorem 10.3 involves the eigenvalues of $\mathbf{A}t$; these are t times the eigenvalues of $\mathbf{A}$. When computing the various derivatives in 10.8, one first calculates the appropriate derivatives of the expression 10.6 with respect to λ, and then substitutes $\lambda = \lambda_i$. The reverse procedure of first substituting $\lambda = \lambda_i$ (a function of t) into 10.6, and then calculating the derivatives with respect to t can give erroneous results.

Example 10.3. Let **A** have four rows and four columns and let $\lambda = 5t$ and $\lambda = 2t$ be eigenvalues of $\mathbf{A}t$ of multiplicities three and one, respectively. Then $n = 4$ and

$$r(\lambda) = \alpha_3 \lambda^3 + \alpha_2 \lambda^2 + \alpha_1 \lambda + \alpha_0$$
$$r'(\lambda) = 3\alpha_3 \lambda^2 + 2\alpha_2 \lambda + \alpha_1$$
$$r''(\lambda) = 6\alpha_3 \lambda + 2\alpha_2$$

Since $\lambda = 5t$ is an eigenvalue of multiplicity three, it follows that $e^{5t} = r(5t)$, $e^{5t} = r'(5t)$ and $e^{5t} = r''(5t)$. Thus,

$$e^{5t} = \alpha_3 (5t)^3 + \alpha_2 (5t)^2 + \alpha_1 (5t) + \alpha_0$$
$$e^{5t} = 3\alpha_3 (5t)^2 + 2\alpha_2 (5t) + \alpha_1$$
$$e^{5t} = 6\alpha_3 (5t) + 2\alpha_2$$

Also, since $\lambda = 2t$ is an eigenvalue of multiplicity one, it follows that $e^{2t} = r(2t)$, or

$$e^{2t} = \alpha_3 (2t)^3 + \alpha_2 (2t)^2 + \alpha_1 (2t) + \alpha_0$$

Notice that we now have four equations in four unknown α's.

Method of computation: For each eigenvalue λ_i of $\mathbf{A}t$, apply Theorem 10.3 to obtain a set of linear equations. When this is done for each eigenvalue, the set of all equations so obtained can be solved for $\alpha_0, \alpha_1,..., \alpha_{n-1}$. These values are then substituted into Equation 10.3 which, in turn, is used to compute $e^{\mathbf{A}t}$.

Solved Problems

Solved Problem 10.1 Find **AB** and **BA** for

$$\mathbf{A} = \begin{bmatrix} 1 & 2 & 3 \\ 4 & 5 & 6 \end{bmatrix}, \qquad \mathbf{B} = \begin{bmatrix} 7 & 0 \\ 8 & -1 \end{bmatrix}.$$

Since **A** has three columns and **B** has two rows, the product **AB** is not defined. But

$$\mathbf{BA} = \begin{bmatrix} 7 & 0 \\ 8 & -1 \end{bmatrix} \begin{bmatrix} 1 & 2 & 3 \\ 4 & 5 & 6 \end{bmatrix}$$

$$= \begin{bmatrix} 7(1)+(0)(4) & 7(2)+(0)(5) & 7(3)+(0)(6) \\ 8(1)+(-1)(4) & 8(2)+(-1)(5) & 8(3)+(-1)(6) \end{bmatrix}$$

$$= \begin{bmatrix} 7 & 14 & 21 \\ 4 & 11 & 18 \end{bmatrix}$$

Solved Problem 10.2 Find the eigenvalues of $\mathbf{A} = \begin{bmatrix} 1 & 3 \\ 4 & 2 \end{bmatrix}$.

We have

$$\mathbf{A} - \lambda\mathbf{I} = \begin{bmatrix} 1 & 3 \\ 4 & 2 \end{bmatrix} + (-\lambda)\begin{bmatrix} 1 & 0 \\ 0 & 1 \end{bmatrix}$$

$$= \begin{bmatrix} 1 & 3 \\ 4 & 2 \end{bmatrix} + \begin{bmatrix} -\lambda & 0 \\ 0 & -\lambda \end{bmatrix} = \begin{bmatrix} 1-\lambda & 3 \\ 4 & 2-\lambda \end{bmatrix}$$

Hence,
$$\det(\mathbf{A} - \lambda\mathbf{I}) = \det\begin{bmatrix} 1-\lambda & 3 \\ 4 & 2-\lambda \end{bmatrix}$$

$$= (1-\lambda)(2-\lambda) - (3)(4) = \lambda^2 - 3\lambda - 10$$

The characteristic equation of $\mathbf{A}$ is $\lambda^2 - 3\lambda - 10 = 0$, which can be factored into $(\lambda - 5)(\lambda + 2) = 0$. The roots of the equation are $\lambda_1 = 5$ and $\lambda_2 = -2$, which are the eigenvalues of $\mathbf{A}$.

Solved Problem 10.3 Find $e^{\mathbf{A}t}$ for $\mathbf{A} = \begin{bmatrix} 1 & 1 \\ 9 & 1 \end{bmatrix}$.

Here $n = 2$. From Equation 10.4,

$$e^{\mathbf{A}t} = \alpha_1 \mathbf{A}t + \alpha_0 \mathbf{I} = \begin{bmatrix} \alpha_1 t + \alpha_0 & \alpha_1 t \\ 9\alpha_1 t & \alpha_1 t + \alpha_0 \end{bmatrix} \qquad (10.9)$$

and from Equation 10.6, $r(\lambda) = \alpha_1\lambda + \alpha_0$. The eigenvalues of $\mathbf{A}t$ are $\lambda_1 = 4t$ and $\lambda_2 = -2t$, which are both of multiplicity one. Substituting these values successively into Equation 10.7, we obtain the two equations

$$e^{4t} = 4t\alpha_1 + \alpha_0$$
$$e^{-2t} = -2t\alpha_1 + \alpha_0$$

Solving these equations for α_1 and α_0, we find that

$$\alpha_1 = \frac{1}{6t}(e^{4t} - e^{-2t}) \quad \text{and} \quad \alpha_0 = \frac{1}{3}(e^{4t} + 2e^{-2t})$$

Substituting these values into 10.9 and simplifying, we have

$$e^{\mathbf{A}t} = \frac{1}{6}\begin{bmatrix} 3e^{4t} + 3e^{-2t} & e^{4t} - e^{-2t} \\ 9e^{4t} - 9e^{-2t} & 3e^{4t} + 3e^{-2t} \end{bmatrix}$$

Solved Problem 10.4 Find $e^{\mathbf{A}t}$ for $\mathbf{A} = \begin{bmatrix} 0 & 1 \\ 8 & -2 \end{bmatrix}$.

Since $n = 2$, it follows from Equations 10.4 and 10.6 that

$$e^{\mathbf{A}t} = \alpha_1 \mathbf{A}t + \alpha_0 \mathbf{I} = \begin{bmatrix} \alpha_0 & \alpha_1 t \\ 8\alpha_1 t & -2\alpha_1 t + \alpha_0 \end{bmatrix} \quad (10.10)$$

and $r(\lambda) = \alpha_1 \lambda + \alpha_0$. The eigenvalues of $\mathbf{A}t$ are $\lambda_1 = 2t$ and $\lambda_2 = -4t$, which are both of multiplicity one. Substituting these values successively into Equation 10.7, we obtain

$$e^{2t} = \alpha_1(2t) + \alpha_0$$
$$e^{-4t} = \alpha_1(-4t) + \alpha_0$$

Solving these equations for α_1 and α_0, we find that

$$\alpha_1 = \frac{1}{6t}(e^{2t} - e^{-4t}) \quad \text{and} \quad \alpha_0 = \frac{1}{3}(2e^{2t} + e^{-4t})$$

Substituting these values into 10.10 and simplifying, we have

$$e^{\mathbf{A}t} = \frac{1}{6}\begin{bmatrix} 4e^{2t} + 2e^{-4t} & e^{2t} - e^{-4t} \\ 8e^{2t} - 8e^{-4t} & 2e^{2t} + 4e^{-4t} \end{bmatrix}$$

Chapter 11

SOLUTIONS OF LINEAR DIFFERENTIAL EQUATIONS WITH CONSTANT COEFFICIENTS BY MATRIX METHODS

IN THIS CHAPTER:

- ✔ *Reduction of Linear Differential Equations to a First-Order System*
- ✔ *Solution of the Initial-Value Problem*
- ✔ *Solution with No Initial Conditions*
- ✔ *Solved Problems*

Reduction of Linear Differential Equations to a First-Order System

Reduction of One Equation

Every initial-value problem of the form

$$b_n(t)\frac{d^n x}{dt^n} + b_{n-1}(t)\frac{d^{n-1}x}{dt^{n-1}} + \cdots + b_1(t)\dot{x} + b_0(t)x = g(t); \quad (11.1)$$

$$x(t_0) = c_0, \quad \dot{x}(t_0) = c_1, \ldots, x^{(n-1)}(t_0) = c_{n-1} \quad (11.2)$$

with $b_n(t) \neq 0$, can be reduced to the first-order matrix system

$$\dot{\mathbf{x}}(t) = \mathbf{A}(t)\mathbf{x}(t) + \mathbf{f}(t)$$
$$\mathbf{x}(t_0) = \mathbf{c} \quad (11.3)$$

where $\mathbf{A}(t)$, $\mathbf{f}(t)$, $\mathbf{c}$, and the initial time t_0 are known. The method of reduction is as follows.

Step 1. Rewrite 11.1 so that $d^n x/dt^n$ appears by itself. Thus,

$$\frac{d^n x}{dt^n} = a_{n-1}(t)\frac{d^{n-1}x}{dt^{n-1}}$$
$$+ \cdots + a_1(t)\dot{x} + a_0(t)x + f(t) \quad (11.4)$$

where $a_j(t) = -b_j(t)/b_n(t)$ ($j = 0,1,\ldots, n-1$) and $f(t) = g(t)/b_n(t)$.

Step 2. Define n new variables (the same number as the order of the original differential equation), $x_1(t)$, $x_2(t),\ldots, x_n(t)$, by the equations

$$x_1(t) = x(t), \quad x_2(t) = \frac{dx(t)}{dt},$$
$$x_3(t) = \frac{d^2 x(t)}{dt^2}, \ldots, x_n(t) = \frac{d^{n-1}x(t)}{dt^{n-1}} \quad (11.5)$$

These new variables are interrelated by the equations

$$\dot{x}_1(t) = x_2(t)$$
$$\dot{x}_2(t) = x_3(t)$$
$$\dot{x}_3(t) = x_4(t) \qquad (11.6)$$
$$\cdots\cdots\cdots\cdots$$
$$\dot{x}_{n-1}(t) = x_n(t)$$

Step 3. Express d_n^x/dt in terms of the new variables. Proceed by first differentiating the last equation of 11.5 to obtain

$$\dot{x}_n(t) = \frac{d}{dt}\left[\frac{d^{n-1}x(t)}{dt^{n-1}}\right] = \frac{d^n x(t)}{dt^n}$$

Then, from Equations 11.4 and 11.5,

$$\begin{aligned}\dot{x}_n(t) &= a_{n-1}(t)\frac{d^{n-1}x(t)}{dt^{n-1}} + \cdots + a_1(t)\dot{x}(t) + a_0(t)x(t) + f(t)\\ &= a_{n-1}(t)x_n(t) + \cdots + a_1(t)x_2(t) + a_0(t)x_1(t) + f(t)\end{aligned}$$

For convenience, we rewrite this last equation so that $x_1(t)$ appears before $x_2(t)$, etc. Thus,

$$\dot{x}_n(t) = a_0(t)x_1(t) + a_1(t)x_2(t) + \cdots + a_{n-1}(t)x_n(t) + f(t) \quad (11.7)$$

Step 4. Equations 11.6 and 11.7 are a system of first-order linear differential equations in $x_1(t)$, $x_2(t)$,..., $x_n(t)$. This system is equivalent to the single matrix equation $\dot{\mathbf{x}}(t) = \mathbf{A}(t)\mathbf{x}(t) + \mathbf{f}(t)$ if we define

$$\mathbf{x}(t) \equiv \begin{bmatrix} x_1(t) \\ x_2(t) \\ \vdots \\ x_n(t) \end{bmatrix} \qquad (11.8)$$

$$\mathbf{f}(t) \equiv \begin{bmatrix} 0 \\ 0 \\ \vdots \\ 0 \\ f(t) \end{bmatrix} \qquad (11.9)$$

$$\mathbf{A}(t) \equiv \begin{bmatrix} 0 & 1 & 0 & 0 & \cdots & 0 \\ 0 & 0 & 1 & 0 & \cdots & 0 \\ 0 & 0 & 0 & 1 & \cdots & 0 \\ \vdots & \vdots & \vdots & \vdots & & \vdots \\ 0 & 0 & 0 & 0 & \cdots & 1 \\ a_0(t) & a_1(t) & a_2(t) & a_3(t) & & a_{n-1}(t) \end{bmatrix} \qquad (11.10)$$

Step 5. Define

$$\mathbf{c} = \begin{bmatrix} c_0 \\ c_1 \\ \vdots \\ c_{n-1} \end{bmatrix}$$

Then the initial conditions (11.2) can be given by the matrix (vector) equation $\mathbf{x}(t_0) = \mathbf{c}$. This last equation is an immediate consequence of Equations 11.8, 11.5, and 11.2, since

$$\mathbf{x}(t_0) = \begin{bmatrix} x_1(t_0) \\ x_2(t_0) \\ \vdots \\ x_n(t_0) \end{bmatrix} = \begin{bmatrix} x(t_0) \\ \dot{x}(t_0) \\ \vdots \\ x^{(n-1)}(t_0) \end{bmatrix} = \begin{bmatrix} c_0 \\ c_1 \\ \vdots \\ c_{n-1} \end{bmatrix} \equiv \mathbf{c}$$

Observe that if no initial conditions are prescribed, Steps 1 through 4 by themselves reduce any linear differential equation 11.1 to the matrix equation $\dot{\mathbf{x}}(t) = \mathbf{A}(t)\mathbf{x}(t) + \mathbf{f}(t)$.

Reduction of a System

A set of linear differential equations with initial conditions also can be reduced to system 11.3. The procedure is nearly identical to the method for reducing a single equation to matrix form; only Step 2 changes. With a system of equations, Step 2 is generalized so that new variables are defined for *each* of the unknown functions in the set.

Solution of the Initial-Value Problem

By the procedure described above, any initial-value problem in which the differential equations are all linear *with constant coefficients*, can be reduced to the matrix system

$$\dot{x}(t) = Ax(t) + f(t); \quad x(t_0) = c \tag{11.11}$$

where A is the matrix of *constants*. The solution to Equation 11.11 is

$$x(t) = e^{A(t-t_0)}c + e^{At} \int_{t_0}^{t} e^{-As}f(s)ds \tag{11.12}$$

or equivalently

$$x(t) = e^{A(t-t_0)}c + \int_{t_0}^{t} e^{A(t-s)}f(s)ds \tag{11.13}$$

In particular, if the initial-value problem is *homogeneous* [i.e., $f(t) = 0$], then both equations 11.12 and 11.13 reduce to

$$x(t) = e^{A(t-t_0)}c \tag{11.14}$$

In the above solutions, the matrices $e^{A(t-t_0)}$, e^{-As}, and $e^{A(t-s)}$ are easily computed from e^{At} by replacing the variable t by $t - t_0$, $-s$, and $t - s$, respectively. Usually $x(t)$ is obtained quicker from 11.13 than from 11.12, since the former equation involves one less matrix multiplication. However, the integrals arising in 11.13 are generally more difficult to evaluate than those in 11.12.

Solution with No Initial Conditions

If no initial conditions are prescribed, the solution of $\dot{x}(t) = Ax(t) + f(t)$ is

$$x(t) = e^{At}k + e^{At} \int e^{-At}f(t)dt \tag{11.15}$$

or, when $f(t) = 0$,

$$\mathbf{x}(t) = e^{\mathbf{A}t}\mathbf{k} \tag{11.16}$$

where $\mathbf{k}$ is an arbitrary constant vector. All constants of integration can be disregarded when computing the integral in Equation 11.15, since they are already included in $\mathbf{k}$.

Solved Problems

Solved Problem 11.1 Put the initial-value problem

$$\ddot{x} + 2\dot{x} - 8x = 0; \quad x(1) = 2, \quad \dot{x}(1) = 3$$

into the form of System 11.3.
Following Step 1, we write $\ddot{x} = -2\dot{x} + 8x$; hence $a_1(t) = -2$, $a_0(t) = 8$, and $f(t) = 0$. Then, defining $x_1(t) = x$ and $x_2(t) = \dot{x}$ (the differential equation is second-order, so we need two new variables), we obtain $\dot{x}_1 = x_2$. Following Step 3, we find

$$\dot{x}_2 = \frac{d^2x}{dt^2} = -2\dot{x} + 8x = -2x_2 + 8x_1$$

Thus,

$$\dot{x}_1 = 0x_1 + 1x_2$$
$$\dot{x}_2 = 8x_1 - 2x_2$$

These equations are equivalent to the matrix equation $\dot{\mathbf{x}}(t) = \mathbf{A}(t)\mathbf{x}(t) + \mathbf{f}(t)$ if we define

$$\mathbf{x}(t) \equiv \begin{bmatrix} x_1(t) \\ x_2(t) \end{bmatrix} \quad \mathbf{A}(t) \equiv \begin{bmatrix} 0 & 1 \\ 8 & -2 \end{bmatrix} \quad \mathbf{f}(t) \equiv \begin{bmatrix} 0 \\ 0 \end{bmatrix}$$

The differential equation is then equivalent to the matrix equation $\dot{\mathbf{x}}(t) = \mathbf{A}(t)\mathbf{x}(t) + \mathbf{f}(t)$, or simply $\dot{\mathbf{x}}(t) = \mathbf{A}(t)\mathbf{x}(t)$, since $\mathbf{f}(t) = 0$. The initial conditions can be given by $\mathbf{x}(t_0) = \mathbf{c}$, if we define $t_0 = 1$ and $\mathbf{c} = \begin{bmatrix} 2 \\ 3 \end{bmatrix}$.

Solved Problem 11.2 Solve $\ddot{x} + 2\dot{x} - 8x = 0$; $x(1) = 2$, $\dot{x}(1) = 3$.

From Problem 11.1, this initial-value problem is equivalent to Equation 11.11 with

$$\mathbf{x}(t) \equiv \begin{bmatrix} x_1(t) \\ x_2(t) \end{bmatrix} \quad \mathbf{A}(t) \equiv \begin{bmatrix} 0 & 1 \\ 8 & -2 \end{bmatrix} \quad \mathbf{f}(t) = \mathbf{0} \quad \mathbf{c} = \begin{bmatrix} 2 \\ 3 \end{bmatrix} \quad t_0 = 1$$

The solution to this system is given by Equation 11.14. For this $\mathbf{A}$, $e^{\mathbf{A}t}$ is given in Problem 10.4; hence,

$$e^{\mathbf{A}(t-t_0)} = e^{\mathbf{A}(t-1)} = \frac{1}{6} \begin{bmatrix} 4e^{2(t-1)} + 2e^{-4(t-1)} & e^{2(t-1)} - e^{-4(t-1)} \\ 8e^{2(t-1)} - 8e^{-4(t-1)} & 2e^{2(t-1)} + 4e^{-4(t-1)} \end{bmatrix}$$

Therefore,

$$\begin{aligned}
\mathbf{x}(t) &= e^{\mathbf{A}(t-1)}\mathbf{c} \\
&= \frac{1}{6} \begin{bmatrix} 4e^{2(t-1)} + 2e^{-4(t-1)} & e^{2(t-1)} - e^{-4(t-1)} \\ 8e^{2(t-1)} - 8e^{-4(t-1)} & 2e^{2(t-1)} + 4e^{-4(t-1)} \end{bmatrix} \begin{bmatrix} 2 \\ 3 \end{bmatrix} \\
&= \frac{1}{6} \begin{bmatrix} 2(4e^{2(t-1)} + 2e^{-4(t-1)}) + 3(e^{2(t-1)} - e^{-4(t-1)}) \\ 2(8e^{2(t-1)} - 8e^{-4(t-1)}) + 3(2e^{2(t-1)} + 4e^{-4(t-1)}) \end{bmatrix} \\
&= \begin{bmatrix} \dfrac{11}{6}e^{2(t-1)} + \dfrac{1}{6}e^{-4(t-1)} \\ \dfrac{22}{6}e^{2(t-1)} - \dfrac{4}{6}e^{-4(t-1)} \end{bmatrix}
\end{aligned}$$

and the solution to the original initial-value problem is

$$x(t) = x_1(t) = \frac{11}{6}e^{2(t-1)} + \frac{1}{6}e^{-4(t-1)}$$

Chapter 12
POWER SERIES SOLUTIONS

IN THIS CHAPTER:

- ✔ *Second-Order Linear Equations with Variable Coefficients*
- ✔ *Analytic Functions and Ordinary Points*
- ✔ *Solutions Around the Origin of Homogenous Equations*
- ✔ *Solutions Around the Origin of Nonhomogeneous Equations*
- ✔ *Initial-Value Problems*
- ✔ *Solutions Around Other Points*
- ✔ *Regular Singular Points*
- ✔ *The Method of Frobenius*
- ✔ *Solved Problems*

Second-Order Linear Equations with Variable Coefficients

A *second-order* linear differential equation

$$b_2(x)y'' + b_1(x)y' + b_0(x)y = g(x) \tag{12.1}$$

has variable coefficients when $b_2(x)$, $b_1(x)$, and $b_0(x)$ are *not* all constants or constant multiples of one another. If $b_2(x)$ is not zero in a given interval, then we can divide by it and rewrite Equation 12.1 as

$$y'' + P(x)y' + Q(x)y = \phi(x) \tag{12.2}$$

where $P(x) = b_1(x)/b_2(x)$, $Q(x) = b_0(x)/b_2(x)$, and $\phi(x) = g(x)/b_2(x)$. In this chapter, we describe procedures for solving many equations in the form of 12.1 and 12.2. These procedures can be generalized in a straightforward manner to solve higher-order linear differential equations with variable coefficients.

Analytic Functions and Ordinary Points

A function $f(x)$ is *analytic* at x_0 if its Taylor series about x_0,

$$\sum_{n=0}^{\infty} \frac{f^{(n)}(x_0)(x - x_0)^n}{n!}$$

converges to $f(x)$ in some neighborhood of x_0.

You Need to Know!

Polynomials, $\sin x$, $\cos x$, and e^x are analytic everywhere.

Sums, differences, and products of polynomials, $\sin x$, $\cos x$, and e^x are also analytic everywhere. Quotients of any two of these functions are analytic at all points where the denominator is not zero.

The point x_0 is an *ordinary point* of the differential equation 12.2 if both $P(x)$ and $Q(x)$ are analytic at x_0. If either of these functions is not analytic at x_0, then x_0 is a *singular point* of 12.2.

Solutions Around the Origin of Homogeneous Equations

Equation 12.1 is *homogeneous* when $g(x) \equiv 0$, in which case Equation 12.2 specializes to

$$y'' + P(x)y' + Q(x)y = 0 \tag{12.3}$$

Theorem 12.1. If $x = 0$ is an ordinary point of Equation 12.3, then the general solution in an interval containing this point has the form

$$y = \sum_{n=0}^{\infty} a_n x^n = a_0 y_1(x) + a_1 y_2(x) \tag{12.4}$$

where a_0 and a_1 are arbitrary constants and $y_1(x)$ and $y_2(x)$ are linearly independent functions analytic at $x = 0$.

To evaluate the coefficients a_n in the solution furnished by Theorem 12.1, use the following five-step procedure known as the *power series method*.

Step 1. Substitute into the left side of the homogeneous differential equation the power series

$$y = \sum_{n=0}^{\infty} a_n x^n = \ a_0 + a_1 x + a_2 x^2 + a_3 x^3 + a_4 x^4 + \cdots \\ +a_n x^n + a_{n+1} x^{n+1} + a_{n+2} x^{n+2} + \cdots \tag{12.5}$$

together with the power series for

$$y' = \ a_1 + 2a_2 x + 3a_3 x^2 + 4a_4 x^3 + \cdots \\ +na_n x^{n-1} + (n+1)a_{n+1} x^n + (n+2)a_{n+2} x^{n+1} + \cdots \tag{12.6}$$

and

$$
\begin{aligned}
y'' = \ & 2a_2 + 6a_3 x + 12a_4 x^2 + \cdots \\
& + n(n-1)a_n x^{n-2} + (n+1)(n)a_{n+1} x^{n-1} \qquad (12.7) \\
& + (n+2)(n+1)a_{n+2} x^n + \cdots
\end{aligned}
$$

Step 2. Collect powers of x and set the coefficients of each power of x equal to zero.

Step 3. The equation obtained by setting the coefficient of x^n to zero in Step 2 will contain a_j terms for a finite number of j values. Solve this equation for the a_j term having the largest subscript. The resulting equation is known as the *recurrence formula* for the given differential equation.

Step 4. Use the recurrence formula to sequentially determine $a_j (j = 2,3,4,...)$ in terms of a_0 and a_1.

Step 5. Substitute the coefficients determined in Step 4 into Equation 12.5 and rewrite the solution in the form of Equation 12.4.

The power series method is only applicable when $x = 0$ is an ordinary point. Although the differential equation must be in the form of Equation 12.2 to determine whether $x = 0$ is an ordinary point, once this condition is verified, the power series method can be used on either form 12.1 or 12.2. If $P(x)$ or $Q(x)$ in 12.2 are quotients of polynomials, it is often simpler first to multiply through by the lowest common denominator, thereby clearing fractions, and then to apply the power series method to the resulting equation in the form of Equation 12.1.

Solutions Around the Origin of Nonhomogeneous Equations

If $\phi(x)$ in Equation 12.2 is analytic at $x = 0$, it has a Taylor series expansion around that point and the power series method given above can be modified to solve either Equation 12.1 or 12.2. In Step 1, Equations 12.5 through 12.7 are substituted into the left side of the nonhomogeneous equation; the right side is written as a Taylor series around the origin.

Steps 2 and 3 change so that the coefficients of each power of x on the left side of the equation resulting from Step 1 are set equal to their counterparts on the right side of that equation. The form of the solution in Step 5 becomes

$$y = a_0 y_1(x) + a_1 y_2(x) + y_3(x)$$

which has the form specified in Theorem 4.4. The first two terms comprise the general solution to the associated homogeneous differential equation while the last function is a particular solution to the nonhomogeneous equation.

Initial-Value Problems

⭐ Important!

Solutions to initial-value problems are obtained by first solving the given differential equation and then applying the specified initial conditions.

Solutions Around Other Points

When solutions are required around the ordinary point $x_0 \neq 0$, it generally simplifies the algebra if x_0 is translated to the origin by the change of variables $t = x - x_0$. The solution of the new differential equation that results can be obtained by the power series method about $t = 0$. Then the solution of the original equation is easily gotten by back-substitution.

Regular Singular Points

The point x_0 is a *regular singular point* of the second-order homogeneous linear differential equation

$$y'' + P(x)y' + Q(x)y = 0 \qquad (12.8)$$

if x_0 is not an ordinary point but both $(x - x_0)P(x)$ and $(x - x_0)^2 Q(x)$ are analytic at x_0. We only consider regular singular points at $x_0 = 0$; if this is not the case, then the change of variables $t = x - x_0$ will translate x_0 to the origin.

Method of Frobenius

Theorem 12.2. If $x_0 = 0$ is a regular singular point of 12.8, then the equation has at least one solution of the form

$$y = x^\lambda \sum_{n=0}^{\infty} a_n x^n$$

where λ and $a_n (n = 0,1,2,...)$ are constants. This solution is valid in an interval $0 < x < R$ for some real number R.

To evaluate the coefficients a_n and λ in Theorem 12.2, one proceeds as in the power series method described above. The infinite series

$$\begin{aligned} y \quad &= x^\lambda \sum_{n=0}^{\infty} a_n x^n = \sum_{n=0}^{\infty} a_n x^{\lambda+n} \\ &= a_0 x^\lambda + a_1 x^{\lambda+1} + a_2 x^{\lambda+2} + \cdots \\ &\quad + a_{n-1} x^{\lambda+n-1} + a_n x^{\lambda+n} + a_{n+1} x^{\lambda+n+1} + \cdots \end{aligned} \tag{12.9}$$

with its derivatives

$$\begin{aligned} y' = \lambda a_0 x^{\lambda-1} + (\lambda+1)a_1 x^\lambda + (\lambda+2)a_2 x^{\lambda+1} + \cdots \\ + (\lambda+n-1)a_{n-1} x^{\lambda+n-2} + (\lambda+n)a_n x^{\lambda+n-1} \\ + (\lambda+n+1)a_{n+1} x^{\lambda+n} + \cdots \end{aligned} \tag{12.10}$$

and

$$\begin{aligned} y'' = \lambda(\lambda-1)a_0 x^{\lambda-2} + (\lambda+1)(\lambda)a_1 x^{\lambda-1} + (\lambda+2)(\lambda+1)a_2 x^\lambda \\ + \cdots + (\lambda+n-1)(\lambda+n-2)a_{n-1} x^{\lambda+n-3} \\ + (\lambda+n)(\lambda+n-1)a_n x^{\lambda+n-2} + (\lambda+n+1)(\lambda+n)a_{n+1} x^{\lambda+n-1} + \cdots \end{aligned} \tag{12.11}$$

are substituted into Equation 12.8. Terms with like powers of x are collected together and set equal to zero. When this is done for x^n the resulting equation is a recurrence formula. A quadratic equation in λ, called the *indicial equation*, arises when the coefficient of x^0 is set to zero and a_0 is left arbitrary.

The two roots of the indicial equation can be real or complex. If complex they will occur in a conjugate pair and the complex solutions that they produce can be combined (by using Euler's relations and the identity $x^{a\pm ib} = x^a e^{\pm ib \ln x}$) to form real solutions. In this book we shall, for simplicity, suppose that both roots of the indicial equation are real. Then, if λ is taken as the *larger* indicial root, $\lambda = \lambda_1 \geq \lambda_2$, the method of Frobenius always yields a solution

$$y_1(x) = x^{\lambda_1} \sum_{n=0}^{\infty} a_n(\lambda_1)x^n \qquad (12.12)$$

to Equation 12.8. [We have written $a_n(\lambda_1)$ to indicate the coefficients produced by the method when $\lambda = \lambda_1$.]

If $P(x)$ and $Q(x)$ are quotients of polynomials, it is usually easier first to multiply 12.8 by their lowest common denominator and then to apply the method of Frobenius to the resulting equation.

General Solution

The method of Frobenius always yields one solution to Equation 12.8 of the form 12.12. The general solution (see Theorem 4.2) has the form $y = c_1 y_1(x) + c_2 y_2(x)$ where c_1 and c_2 are arbitrary constants and $y_2(x)$ is a second solution of 12.8 that is linearly independent from $y_1(x)$. The method for obtaining this second solution depends on the relationship between the two roots of the indicial equation.

Case 1. If $\lambda_1 - \lambda_2$, is not an integer, then

$$y_2(x) = x^{\lambda_2} \sum_{n=0}^{\infty} a_n(\lambda_2)x^n \qquad (12.13)$$

where $y_2(x)$ is obtained in an identical manner as $y_1(x)$ by the method of Frobenius, using λ_2 in place of λ_1.

Case 2. If, $\lambda_1 = \lambda_2$, then

$$y_2(x) = y_1(x)\ln x + x^{\lambda_2} \sum_{n=0}^{\infty} b_n(\lambda_1)x^n \qquad (12.14)$$

To generate this solution, keep the recurrence formula in terms of λ and use it to find the coefficients a_n ($n \geq 1$) in terms of both λ and a_0, where the coefficient a_0 remains arbitrary. Substitute these a_n into Equation 12.9 to obtain a function $y(\lambda, x)$ which depends on the variables λ and x. Then

$$y_2(x) = \frac{\partial y(\lambda, x)}{\partial \lambda}\bigg|_{\lambda=\lambda_1} \tag{12.15}$$

Case 3. If $\lambda_1 - \lambda_2 = N$, a positive integer, then

$$y_2(x) = d_{-1}y_1(x)\ln x + x^{\lambda_2} \sum_{n=0}^{\infty} d_n(\lambda_2)x^n \tag{12.16}$$

To generate this solution, first try the method of Frobenius with λ_2. If it yields a second solution, then this solution is $y_2(x)$, having the form of 12.16 with $d_{-1} = 0$. Otherwise, proceed as in Case 2 to generate $y(\lambda, x)$, whence

$$y_2(x) = \frac{\partial}{\partial x}[(\lambda - \lambda_2)y(\lambda, x)]\big|_{\lambda=\lambda_2} \tag{12.17}$$

Solved Problems

Solved Problem 12.1 Determine whether $x = 0$ is an ordinary point of the differential equation

$$y'' - xy' + 2y = 0$$

Here $P(x) = -x$ and $Q(x) = 2$ are both polynomials; hence they are analytic everywhere. Therefore, every value of x, in particular $x = 0$, is an ordinary point.

Solved Problem 12.2 Find a recurrence formula for the power series solution around $x = 0$ for the differential equation given in Problem 12.1.

It follows from Problem 12.1 that $x = 0$ is an ordinary point of the given equation, so Theorem 12.1 holds. Substituting Equations 12.5 through 12.7 into the left side of the differential equation, we find

$$[2a_2 + 6a_3x + 12a_4x^2 + \cdots + n(n-1)a_nx^{n-2} + (n+1)(n)a_{n+1}x^{n-1} + \cdots]$$
$$-x[a_1 + 2a_2x + 3a_3x^2 + \cdots + na_nx^{n-1} + (n+1)a_{n+1}x^n + \cdots]$$
$$+2[a_0 + a_1x + a_2x^2 + a_3x^3 + \cdots + a_nx^n + a_{n+1}x^{n+1} + \cdots] = 0$$

Combining terms that contain like powers of x, we have

$$(2a_2 + 2a_0) + x(6a_3 + a_1) + x^2(12a_4) + x^3(20a_5 - a_3)$$
$$+\cdots + x^n[(n+2)(n+1)a_{n+2} - na_n + 2a_n] + \cdots$$
$$= 0 + 0x + 0x^2 + 0x^3 + \cdots + 0x^n + \cdots$$

The last equation holds if and only if each coefficient in the left-hand side is zero. Thus,

$$2a_2 + 2a_0 = 0, \quad 6a_3 + a_1 = 0, \quad 12a_4 = 0, \quad 20a_5 - a_3 = 0, \ldots$$

In general,

$$(n+2)(n+1)a_{n+2} - (n-2)a_n = 0, \quad \text{or,}$$
$$a_{n+2} = \frac{(n-2)}{(n+2)(n+1)}a_n$$

which is the recurrence formula for this problem.

Solved Problem 12.3 Find the general solution near $x = 0$ of $y'' - xy' + 2y = 0$.

Successively evaluating the recurrence formula obtained in Problem 12.2 for $n = 0,1,2,\ldots$, we calculate

$$a_2 = -a_0$$
$$a_3 = -\frac{1}{6}a_1$$
$$a_4 = 0$$
$$a_5 = \frac{1}{20}a_3 = \frac{1}{20}\left(-\frac{1}{6}a_1\right) = -\frac{1}{120}a_1$$

(12.18)

$$a_6 = \frac{2}{30}a_4 = \frac{1}{15}(0)$$

$$a_7 = \frac{3}{42}a_5 = \frac{1}{14}\left(-\frac{1}{120}a_1\right) = -\frac{1}{1680}a_1$$

$$a_8 = \frac{4}{56}a_6 = \frac{1}{14}(0) = 0$$

....

Note that since $a_4 = 0$, it follows from the recurrence formula that all the even coefficients beyond a_4 are also zero. Substituting 12.18 into Equation 12.5 we have

$$y = a_0 + a_1 x - a_0 x^2 - \frac{1}{6}a_1 x^3 + 0x^4 - \frac{1}{120}a_1 x^5 + 0x^6 - \frac{1}{1680}a_1 x^7 - \cdots$$

$$= a_0(1-x^2) + a_1\left(x - \frac{1}{6}x^3 - \frac{1}{120}x^5 - \frac{1}{1680}x^7 - \cdots\right)$$

If we define

$$y_1(x) = 1 - x^2 \quad \text{and} \quad y_2(x) = x - \frac{1}{6}x^3 - \frac{1}{120}x^5 - \frac{1}{1680}x^7 - \cdots$$

then the general solution can be rewritten as $y = a_0 y_1(x) + a_1 y_2(x)$.

Solved Problem 12.4 Determine whether $x = 0$ is a regular singular point of the differential equation

$$8x^2 y'' + 10xy' + (x-1)y = 0$$

Dividing by $8x^2$, we have

$$P(x) = \frac{5}{4x} \quad \text{and} \quad Q(x) = \frac{1}{8x} - \frac{1}{8x^2}$$

Neither of these functions is defined at $x = 0$, so this point is a singular point. Furthermore, both

$$xP(x) = \frac{5}{4} \quad \text{and} \quad x^2 Q(x) = \frac{1}{8}(x-1)$$

are analytic everywhere: the first is a constant and the second a polynomial. Hence, both are analytic at $x = 0$, and this point is a regular singular point.

Solved Problem 12.5 Use the method of Frobenius to find one solution near $x = 0$ of $x^2 y'' - xy' + y = 0$.

Here $P(x) = -1/x$ and $Q(x) = 1/x^2$, so $x = 0$ is a regular singular point and the method of Frobenius is applicable. Substituting Equations 12.9 through 12.11 into the left side of the differential equation, as given and combining coefficients of like powers of x, we obtain

$$x^\lambda (\lambda - 1)^2 a_0 + x^{\lambda+1}[\lambda^2 a_1] + \cdots$$
$$+ x^{\lambda+n}[(\lambda + n)^2 - 2(\lambda + n) + 1]a_n + \cdots = 0$$

Thus,

$$(\lambda - 1)^2 a_0 = 0 \qquad (12.19)$$

and, in general,

$$[(\lambda + n)^2 - 2(\lambda + n) + 1]a_n = 0 \qquad (12.20)$$

From Equation 12.19, the indicial equation is $(\lambda - 1)^2 = 0$, which has roots $\lambda_1 = \lambda_2 = 1$. Substituting $\lambda = 1$ into Equation 12.20 we obtain $n^2 a_n = 0$, which implies that $a_n = 0$, $n \geq 1$. Thus, $y_1(x) = a_0 x$.

Solved Problem 12.6 Find the general solution near $x = 0$ of $3x^2 y'' - xy' + y = 0$.

Here $P(x) = -1/(3x)$ and $Q(x) = 1/(3x^2)$; hence, $x = 0$, is a regular singular point and the method of Frobenius is applicable. Substituting Equations 12.9 through 12.11 into the differential equation and simplifying, we have

$$x^\lambda [3\lambda^2 - 4\lambda + 1]a_0 + x^{\lambda+1}[3\lambda^2 + 2\lambda]a_1 + \cdots$$
$$+ x^{\lambda+n}[3(\lambda + n)^2 - 4(\lambda + n) + 1]a_n + \cdots = 0$$

Dividing by x^λ and equating all coefficients to zero, we find

$$(3\lambda^2 - 4\lambda + 1)a_0 = 0 \tag{12.21}$$

and

$$[3(\lambda + n)^2 - 4(\lambda + n) + 1]a_n = 0 \quad n \geq 1 \tag{12.22}$$

From 12.21, we conclude that the indicial equation is $3\lambda^2 - 4\lambda + 1 = 0$, which has roots $\lambda_1 = 1$ and $\lambda_2 = 1/3$. Since $\lambda_1 - \lambda_2 = 2/3$, the solution is given by Equations 12.12 and 12.13. Note that for either value of λ, Equation 12.22 is satisfied by simply choosing $a_n = 0$, $n \geq 1$. Thus,

$$y_1(x) = x^1 \sum_{n=0}^{\infty} a_n x^n = a_0 x$$

$$y_2(x) = x^{1/3} \sum_{n=0}^{\infty} a_n x^n = a_0 x^{1/3}$$

and the general solution is

$$y = c_1 y_1(x) + c_2 y_2(x) = k_1 x + k_2 x^{1/3}$$

where $k_1 = c_1 a_0$ and $k_2 = c_2 a_0$.

Solved Problem 12.7 Use the method of Frobenius to find one solution near $x = 0$ of $x^2 y'' + (x^2 - 2x)y' + 2y = 0$.

Here $P(x) = 1 - \dfrac{2}{x}$ and $Q(x) = \dfrac{2}{x^2}$; so, $x = 0$ is a regular singular point and the method of Frobenius is applicable. Substituting Equations 12.9 through 12.11 into the left side of the differential equation, as given, and combining coefficients of like powers of x, we obtain

$$x^\lambda[(\lambda^2 - 3\lambda + 2)a_0] + x^{\lambda+1}[(\lambda^2 - \lambda)a_1 + \lambda a_0] + \cdots$$
$$+ x^{\lambda+n}\{[(\lambda + n)^2 - 3(\lambda + n) + 2]a_n + (\lambda + n - 1)a_{n-1}\} + \cdots = 0$$

Dividing by x^λ, factoring the coefficient of a_n, and equating the coefficient of each power of x to zero, we obtain

$$(\lambda^2 - 3\lambda + 2)a_0 = 0 \tag{12.23}$$

and, in general, $[(\lambda+n)-2][(\lambda+n)-1]a_n +(\lambda+n-1)a_{n-1} = 0$, or,

$$a_n = -\frac{1}{\lambda+n-2}a_{n-1} \quad (n \geq 1) \tag{12.24}$$

From 12.23, the indicial equation is $\lambda^2 - 3\lambda + 2 = 0$, which has roots λ_1 = 2 and $\lambda_2 = 1$. Since $\lambda_1 - \lambda_2 = 1$, a positive integer, the solution is given by Equations 12.12 and 12.16. Substituting $\lambda = 2$ into 12.24, we have $a_n = -(1/n)a_{n-1}$, from which we obtain

$$a_1 = -a_0$$
$$a_2 = -\frac{1}{2}a_1 = \frac{1}{2!}a_0$$
$$a_3 = -\frac{1}{3}a_2 = -\frac{1}{3}\frac{1}{2!}a_0 = -\frac{1}{3!}a_0$$

and, in general, $a_k = \frac{(-1)^k}{k!}a_0$. Thus,

$$y_1(x) = a_0 x^2 \sum_{n=0}^{\infty} \frac{(-1)^n}{n!}x^n = a_0 x^2 e^{-x}$$

Chapter 13
GAMMA AND BESSEL FUNCTIONS

IN THIS CHAPTER:

✔ *Gamma Function*
✔ *Bessel Functions*
✔ *Algebraic Operations on Infinite Series*
✔ *Solved Problems*

Gamma Function

The *gamma function*, $\Gamma(p)$, is defined for any positive real number p by

$$\Gamma(p) = \int_0^\infty x^{p-1} e^{-x} dx \qquad (13.1)$$

Consequently, $\Gamma(1) = 1$ and for any positive real number p,

$$\Gamma(p+1) = p\Gamma(p) \qquad (13.2)$$

Furthermore, when $p = n$, a positive integer,

$$\Gamma(n+1) = n! \qquad (13.3)$$

Thus, the gamma function (which is defined on all positive real numbers) is an extension of the factorial function (which is defined only on the non-negative integers).

Equation 13.2 may be rewritten as

$$\Gamma(p) = \frac{1}{p}\Gamma(p+1) \tag{13.4}$$

which defines the gamma function iteratively for all nonintegral negative values of p. $\Gamma(0)$ remains undefined, because

$$\lim_{p \to 0^+} \Gamma(p) = \lim_{p \to 0^+} \frac{\Gamma(p+1)}{p} = \infty$$

and

$$\lim_{p \to 0^-} \Gamma(p) = \lim_{p \to 0^-} \frac{\Gamma(p+1)}{p} = -\infty$$

It then follows from Equation 13.4 that $\Gamma(p)$ is undefined for negative integer values of p.

Table 13.1 lists values of the gamma function in the interval $1 \le p < 2$. These tabular values are used with Equations 13.2 and 13.4 to generate values of $\Gamma(p)$ in other intervals.

Bessel Functions

Let p represent any real number. The *Bessel function of the first kind of order p, $J_p(x)$*, is

$$J_p(x) = \sum_{k=0}^{\infty} \frac{(-1)^k x^{2k+p}}{2^{2k+p} k! \Gamma(p+k+1)} \tag{13.5}$$

The function $J_p(x)$ is a solution near the regular singular point $x = 0$ of *Bessel's differential equation of order p*:

$$x^2 y'' + xy' + (x^2 - p^2)y = 0 \tag{13.6}$$

In fact, $J_p(x)$ is that solution of Equation 13.6 guaranteed by Theorem 12.2.

x	$\Gamma(x)$	x	$\Gamma(x)$	x	$\Gamma(x)$
1.00	1.0000 0000	1.34	0.8922 1551	1.67	0.9032 9650
1.01	0.9943 2585	1.35	0.8911 5144	1.68	0.9050 0103
1.02	0.9888 4420	1.36	0.8901 8453	1.69	0.9067 8182
1.03	0.9835 4995	1.37	0.8893 1351	1.70	0.9086 3873
1.04	0.9784 3820	1.38	0.8885 3715	1.71	0.9105 7168
1.05	0.9735 0427	1.39	0.8878 5429	1.72	0.9125 8058
1.06	0.9687 4365	1.40	0.8872 6382	1.73	0.9146 6537
1.07	0.9641 5204	1.41	0.8867 6466	1.74	0.9168 2603
1.08	0.9597 2531	1.42	0.8863 5579	1.75	0.9190 6253
1.09	0.9554 5949	1.43	0.8860 3624	1.76	0.9213 7488
1.10	0.9513 5077	1.44	0.8858 0506	1.77	0.9237 6313
1.11	0.9473 9550	1.45	0.8856 6138	1.78	0.9262 2731
1.12	0.9435 9019	1.46	0.8856 0434	1.79	0.9287 6749
1.13	0.9399 3145	1.47	0.8856 3312	1.80	0.9313 8377
1.14	0.9364 1607	1.48	0.8857 4696	1.81	0.9340 7626
1.15	0.9330 4093	1.49	0.8859 4513	1.82	0.9368 4508
1.16	0.9298 0307	1.50	0.8862 2693	1.83	0.9396 9040
1.17	0.9266 9961	1.51	0.8865 9169	1.84	0.9426 1236
1.18	0.9237 2781	1.52	0.8870 3878	1.85	0.9456 1118
1.19	0.9208 8504	1.53	0.8875 6763	1.86	0.9486 8704
1.20	0.9181 6874	1.54	0.8881 7766	1.87	0.9518 4019
1.21	0.9155 7649	1.55	0.8888 6835	1.88	0.9550 7085
1.22	0.9131 0595	1.56	0.8896 3920	1.89	0.9583 7931
1.23	0.9107 5486	1.57	0.8904 8975	1.90	0.9617 6583
1.24	0.9085 2106	1.58	0.8914 1955	1.91	0.9652 3073
1.25	0.9064 0248	1.59	0.8924 2821	1.92	0.9687 7431
1.26	0.9043 9712	1.60	0.8935 1535	1.93	0.9723 9692
1.27	0.9025 0306	1.61	0.8946 8061	1.94	0.9760 9891
1.28	0.9007 1848	1.62	0.8959 2367	1.95	0.9798 8065
1.29	0.8990 4159	1.63	0.8972 4423	1.96	0.9837 4254
1.30	0.8974 7070	1.64	0.8986 4203	1.97	0.9876 8498
1.31	0.8960 0418	1.65	0.9001 1682	1.98	0.9917 0841
1.32	0.8946 4046	1.66	0.9016 6837	1.99	0.9958 1326
1.33	0.8933 7805				

Table 13.1

Algebraic Operations on Infinite Series

Changing the dummy index. The dummy index in an infinite series can be changed at will without altering the series. For example,

$$\sum_{k=0}^{\infty} \frac{1}{(k+1)!} = \sum_{n=0}^{\infty} \frac{1}{(n+1)!} = \sum_{p=0}^{\infty} \frac{1}{(p+1)!} = \frac{1}{1!} + \frac{1}{2!} + \frac{1}{3!} + \frac{1}{4!} + \frac{1}{5!} + \cdots$$

Change of variables. Consider the infinite series $\displaystyle\sum_{k=0}^{\infty} \frac{1}{(k+1)!}$. If we make the change of variables $j = k + 1$, or $k = j - 1$, then

$$\sum_{k=0}^{\infty} \frac{1}{(k+1)!} = \sum_{j=1}^{\infty} \frac{1}{j!}$$

Note that a change of variables generally changes the limits on the summation. For instance, if $j = k + 1$, it follows that $j = 1$ when $k = 0$, $j = \infty$, when $k = \infty$, and, as k runs from 0 to ∞, j runs from 1 to ∞.

The two operations given above are often used in concert. For example,

$$\sum_{k=0}^{\infty} \frac{1}{(k+1)!} = \sum_{j=2}^{\infty} \frac{1}{(j-1)!} = \sum_{k=2}^{\infty} \frac{1}{(k-1)!}$$

Here, the second series results from the change of variables $j = k + 2$ in the first series, while the third series is the result of simply changing the dummy index in the second series from j to k. Note that all three series equal

$$\frac{1}{1!} + \frac{1}{2!} + \frac{1}{3!} + \frac{1}{4!} + \cdots = e - 1$$

Solved Problems

Solved Problem 13.1 Prove that $\Gamma(p + 1) = p\Gamma(p)$, $p > 0$.

Using Equation 13.1 and integration by parts, we have

$$\Gamma(p+1) = \int_0^\infty x^{(p+1)-1}e^{-x}dx = \lim_{r\to\infty}\int_0^r x^p e^{-x}dx$$

$$= \lim_{r\to\infty}\left[-x^p e^{-x}\Big|_0^r + \int_0^r px^{p-1}e^{-x}dx\right]$$

$$= \lim_{r\to\infty}(-r^p e^{-r}+0) + p\int_0^\infty x^{p-1}e^{-x}dx = p\Gamma(p)$$

The result $\lim_{r\to\infty} r^p e^{-r} = 0$ is easily obtained by first writing $r^p e^{-r}$ as r^p/e^r and using L'Hôpital's rule.

Solved Problem 13.2 Use the method of Frobenius to find one solution of Bessel's equation of order p:

$$x^2 y'' + xy' + (x^2 - p^2)y = 0$$

Substituting Equations 12.9 through 12.11 into Bessel's equation and simplifying, we find that

$$x^\lambda(\lambda^2 - p^2)a_0 + x^{\lambda+1}[(\lambda+1)^2 - p^2]a_1 +$$
$$+ x^{\lambda+2}\{[(\lambda+2)^2 - p^2]a_2 + a_0\} + \cdots$$
$$+ x^{\lambda+n}\{[(\lambda+n)^2 - p^2]a_n + a_{n-2}\} + \cdots = 0$$

Thus,

$$(\lambda^2 - p^2)a_0 = 0 \quad [(\lambda+1)^2 - p^2]a_1 = 0 \qquad (13.7)$$

and, in general, $[(\lambda+n)^2 - p^2]a_n + a_{n-2} = 0$, or

$$a_n = -\frac{1}{(\lambda+n)^2 - p^2}a_{n-2} \quad (n \geq 2) \qquad (13.8)$$

The indicial equation is $\lambda^2 - p^2 = 0$, which has roots $\lambda_1 = p$ and $\lambda_2 = -p$ (p nonnegative). Substituting $\lambda = p$ into 13.7 and 13.8 and simplifying, we find that $a_1 = 0$ and

$$a_n = -\frac{1}{n(2p+n)}a_{n-2} \quad (n \geq 2)$$

Hence, $0 = a_1 = a_3 = a_5 = a_7 = \cdots$ and

$$a_2 = \frac{-1}{2^2 1!(p+1)} a_0$$

$$a_4 = -\frac{1}{2^2 2(p+2)} a_2 = \frac{1}{2^4 2!(p+2)(p+1)} a_0$$

$$a_6 = -\frac{1}{2^2 3(p+3)} a_4 = \frac{-1}{2^6 3!(p+3)(p+2)(p+1)} a_0$$

and, in general,

$$a_{2k} = \frac{(-1)^k}{2^{2k} k!(p+k)(p+k-1)\cdots(p+2)(p+1)} a_0 \quad (k \geq 1)$$

$$y_1(x) = x^\lambda \sum_{n=0}^{\infty} a_n x^n = x^p \left[a_0 + \sum_{k=1}^{\infty} a_{2k} x^{2k} \right]$$

Thus,

$$= a_0 x^p \left[1 + \sum_{k=1}^{\infty} \frac{(-1)^k x^{2k}}{2^{2k} k!(p+k)(p+k-1)\cdots(p+2)(p+1)} \right]$$

It is customary to choose the arbitrary constant a_0 as $a_0 = \dfrac{1}{2^p \Gamma(p+1)}$.

Then bringing $a_0 x^p$ inside the brackets and summation, combining, and finally using Problem 13.1, we obtain

$$y_1(x) = \frac{1}{2^p \Gamma(p+1)} x^p + \sum_{k=1}^{\infty} \frac{(-1)^k x^{2k+p}}{2^{2k+p} k! \Gamma(p+k+1)}$$

$$= \sum_{k=0}^{\infty} \frac{(-1)^k x^{2k+p}}{2^{2k+p} k! \Gamma(p+k+1)} \equiv J_p(x)$$

Chapter 14

NUMERICAL METHODS FOR FIRST-ORDER DIFFERENTIAL EQUATIONS

In This Chapter:

- ✔ *Direction Fields*
- ✔ *Euler's Method*
- ✔ *General Remarks Regarding Numerical Methods*
- ✔ *Modified Euler's Method*
- ✔ *Runge-Kutta Method*
- ✔ *Adams-Bashforth-Moulton Method*
- ✔ *Milne's Method*
- ✔ *Order of a Numerical Method*
- ✔ *Numerical Methods for Systems*
- ✔ *Solved Problems*

Direction Fields

Graphical methods produce plots of solutions to first-order differential equations of the form

$$y' = f(x, y) \qquad (14.1)$$

where the derivative appears only on the left side of the equation.

Example 14.1 (*a*) For the problem $y' = -y + x + 2$, we have $f(x, y) = -y + x + 2$. (*b*) For the problem $y' = y + 1$, we have $f(x, y) = y^2 + 1$. (*c*) For the problem $y' = 3$, we have $f(x, y) = 3$. Observe that in a particular problem, $f(x, y)$ may be independent of x, of y, or of x and y.

Equation 14.1 defines the slope of the solution curve $y(x)$ at any point (x, y) in the plane. A *line element* is a short line segment that begins at the point (x, y) and has a slope specified by 14.1; it represents an approximation to the solution curve through that point. A collection of line elements is a *direction field*. Note that direction fields are tedious to draw by hand, but they can be easily created by many computer algebra systems. The graphs of solutions to 14.1 are generated from direction fields by drawing curves that pass through the points at which line elements are drawn and also are tangent to those line elements.

Note!

A direction field can be used to graphically determine the behavior of the solution.

If the left side of Equation 14.1 is set equal to a constant, the graph of the resulting equation is called an *isocline*. Different constants define different isoclines, and each has the property that all elements emanating from points on that isocline have the same slope, a slope equal to the constant that generated the isocline. When they are simple to draw, isoclines yield many line elements at once which are useful for constructing direction fields.

Euler's Method

If an initial condition of the form

$$y(x_0) = y_0 \tag{14.2}$$

is also specified, then the only solution curve of Equation 14.1 of interest is the one that passes through the initial point (x_0, y_0).

To obtain a graphical approximation to the solution curve of Equations 14.1 and 14.2, begin by constructing a line element at the initial point (x_0, y_0) and then continuing it for a short distance. Denote the terminal point of this line element as (x_1, y_1). Then construct a second line element at (x_1, y_1) and continue it a short distance. Denote the terminal point of this second line element as (x_2, y_2). Follow with a third line element con-
structed at (x_2, y_2) and continue it a short distance. The process proceeds iteratively and concludes when enough of the solution curve has been drawn to meet the needs of those concerned with the problem.

If the difference between successive x values are equal, that is, if for a specified constant h, $h = x_1 - x_0 = x_2 - x_1 = x_3 - x_2 = \cdots$, then the graphical method given above for a first-order initial-value problem is known as Euler's method. It satisfies the formula

$$y_{n+1} = y_n + hf(x_n, y_n) \tag{14.3}$$

for $n = 1, 2, 3, \ldots$. This formula is often written as

$$y_{n+1} = y_n + hy_n' \tag{14.4}$$

where

$$y_n' = f(x_n, y_n) \tag{14.5}$$

as required by Equation 14.1.

Stability

The constant h in Equations 14.3 and 14.4 is called the *step-size*, and its value is arbitrary. In general, the smaller the step-size, the more accurate the approximate solution becomes at the price of more work to obtain that solution. Thus, the final choice of h may be a compromise between accuracy and effort. If h is chosen too large, then the approximate solution may not resemble the real solution at all, a condition known as *numerical instability*. To avoid numerical instability, Euler's method is repeated, each time with a step-size one-half its previous value, until two successive approximations are close enough to each other to satisfy the needs of the solver.

General Remarks Regarding Numerical Methods

A *numerical method* for solving an initial-value problem is a procedure that produces approximate solutions at particular points using only the operations of addition, subtraction, multiplication, division, and functional evaluations. In this chapter, we consider only first-order initial-value problems of the form

$$y' = f(x, y); \quad y(x_0) = y_0 \tag{14.6}$$

Generalizations to higher-order problems are given later in this chapter. Each numerical method will produce approximate solutions at the points $x_0, x_1, x_2, \ldots$, where the difference between any two successive x-values is a constant step-size h; that is $x_{n+1} - x_n = h$ ($n = 0, 1, 2, \ldots$). Remarks made previously in this chapter on the step-size remain valid for all the numerical methods presented.

The approximate solution at x_n will be designated by $y(x_n)$, or simply y_n. The true solution at x_n will be denoted by either $Y(x_n)$ or Y_n. Note that once y_n is known, Equation 14.6 can be used to obtain y'_n as

$$y'_n = f(x_n, y_n) \tag{14.7}$$

The simplest numerical method is Euler's method, described earlier in this chapter.

A *predictor-corrector* method is a set of two equations for y_{n+1}. The first equation, called the *predictor*, is used to predict (obtain a first approximation to) y_{n+1}; the second equation, called the *corrector*, is then used to obtain a corrected value (second approximation) to y_{n+1}. In general, the corrector depends on the predicted value.

Modified Euler's Method

This is a simple predictor-corrector method that uses Euler's method as the predictor and then uses the average value of y' at both the left and right end points of the interval $[x_n, x_{n+1}]$ ($n = 0,1,2,...$) as the slope of the line element approximation to the solution over that interval. The resulting equations are:

$$\text{predictor:} \quad y_{n+1} = y_n + hy'_n$$
$$\text{corrector:} \quad y_{n+1} = y_n + \frac{h}{2}(y'_{n+1} + y'_n)$$

For notational convenience, we designate the predicted value of y_{n+1} by py_{n+1}. It then follows from Equation 14.7 that

$$py'_{n+1} = f(x_{n+1}, py_{n+1}) \tag{14.8}$$

The modified Euler's method becomes

$$\text{predictor:} \quad py_{n+1} = y_n + hy'_n$$
$$\text{corrector:} \quad y_{n+1} = y_n + \frac{h}{2}(py'_{n+1} + y'_n) \tag{14.9}$$

Runge-Kutta Method

$$y_{n+1} = y_n + \frac{1}{6}(k_1 + 2k_2 + 2k_3 + k_4) \tag{14.10}$$

where

$$k_1 = hf(x_n, y_n)$$
$$k_2 = hf(x_n + \frac{1}{2}h, y_n + \frac{1}{2}k_1)$$
$$k_3 = hf(x_n + \frac{1}{2}h, y_n + \frac{1}{2}k_2)$$
$$k_4 = hf(x_n + h, y_n + k_3)$$

This is *not* a predictor-corrector method.

Adams-Bashforth-Moulton Method

predictor: $py_{n+1} = y_n + \dfrac{h}{24}(55y'_n - 59y'_{n-1} + 37y'_{n-2} - 9y'_{n-3})$

corrector: $y_{n+1} = y_n + \dfrac{h}{24}(9py'_{n+1} + 19y'_n - 5y'_{n-1} + y'_{n-2})$ (14.11)

Milne's Method

predictor: $py_{n+1} = y_{n-3} + \dfrac{4h}{3}(2y'_n - y'_{n-1} + 2y'_{n-2})$

corrector: $y_{n+1} = y_{n-1} + \dfrac{h}{3}(py'_{n+1} + 4y'_n + y'_{n-1})$ (14.12)

Starting Values

The Adams-Bashforth-Moulton method and Milne's method require information at y_0, y_1, y_2, and y_3 to start. The first of these values is given by the initial condition in Equation 14.6. The other three starting values are gotten by the Runge-Kutta method.

Order of a Numerical Method

A numerical method is of *order n*, where n is a positive integer, if the method is exact for polynomials of degree n or less. In other words, if the true solution of an initial-value problem is a polynomial of degree n or less, then the approximate solution and the true solution will be identical for a method of order n.

   In general, the higher the order, the more accurate the method. Euler's method, Equation 14.4, is of order one, the modified Euler's method, Equation 14.9, is of order two, while the other three, Equations 14.10 through 14.12, are fourth-order methods.

Numerical Methods for Systems

First-Order Systems

Numerical methods for solving first-order initial-value problems, including all of those described previously in this chapter, are easily extended to a *system* of first-order initial-value problems. These methods are also applicable to most *higher-order* initial-value problems, in particular those that can be transformed to a system of first-order differential equations by the reduction process described in Chapter Eleven.

Standard form for a system of two equations is

$$y' = f(x, y, z)$$
$$z' = g(x, y, z); \qquad (14.13)$$
$$y(x_0) = y_0, \quad z(x_0) = z_0$$

We note that, with $y' = f(x, y, z) = z$, with system 14.13 represents the second-order initial-value problem

$$y'' = g(x, y, y'); \quad y(x_0) = y_0, \quad y'(x_0) = z_0$$

Standard form for a system of three equations is

$$y' = f(x, y, z, w)$$
$$z' = g(x, y, z, w)$$
$$w' = r(x, y, z, w); \qquad (14.14)$$
$$y(x_0) = y_0, \quad z(x_0) = z_0, \quad w(x_0) = w_0$$

If, in such a system, $f(x, y, z, w) = z$, and $g(x, y, z, w) = w$, then system 14.14 represents the third-order initial-value problem

$$y''' = r(x, y, z, w); \quad y(x_0) = y_0, \quad y'(x_0) = z_0, \quad y''(x_0) = w_0$$

The formulas that follow are for systems of two equations in standard form (14.13). Generalizations to systems of three equations in standard form (14.14) or systems with four or more equations are straightforward.

Euler's Method

$$y_{n+1} = y_n + hy'_n$$
$$z_{n+1} = z_n + hz'_n$$

(14.15)

Runge-Kutta Method

$$y_{n+1} = y_n + \frac{1}{6}(k_1 + 2k_2 + 2k_3 + k_4)$$
$$z_{n+1} = z_n + \frac{1}{6}(l_1 + 2l_2 + 2l_3 + l_4)$$

(14.16)

where

$$k_1 = hf(x_n, y_n, z_n)$$
$$l_1 = hg(x_n, y_n, z_n)$$
$$k_2 = hf(x_n + \tfrac{1}{2}h, y_n + \tfrac{1}{2}k_1, z_n + \tfrac{1}{2}l_1)$$
$$l_2 = hg(x_n + \tfrac{1}{2}h, y_n + \tfrac{1}{2}k_1, z_n + \tfrac{1}{2}l_1)$$
$$k_3 = hf(x_n + \tfrac{1}{2}h, y_n + \tfrac{1}{2}k_2, z_n + \tfrac{1}{2}l_2)$$
$$l_3 = hg(x_n + \tfrac{1}{2}h, y_n + \tfrac{1}{2}k_2, z_n + \tfrac{1}{2}l_2)$$
$$k_4 = hf(x_n + h, y_n + k_3, z_n + l_3)$$
$$l_4 = hg(x_n + h, y_n + k_3, z_n + l_3)$$

Adams-Bashforth-Moulton Method

predictors: $$py_{n+1} = y_n + \frac{h}{24}(55y'_n - 59y'_{n-1} + 37y'_{n-2} - 9y'_{n-3})$$
$$pz_{n+1} = z_n + \frac{h}{24}(55z'_n - 59z'_{n-1} + 37z'_{n-2} - 9z'_{n-3})$$

correctors: $$y_{n+1} = y_n + \frac{h}{24}(9py'_{n+1} + 19y'_n - 5y'_{n-1} + y'_{n-2})$$ (14.17)
$$z_{n+1} = z_n + \frac{h}{24}(9pz'_{n+1} + 19z'_n - 5z'_{n-1} + z'_{n-2})$$

Corresponding derivatives are calculated from system 14.13. In particular,

$$y'_{n+1} = f(x_{n+1}, y_{n+1}, z_{n+1})$$
$$z'_{n+1} = g(x_{n+1}, y_{n+1}, z_{n+1})$$

(14.18)

The derivatives associated with the predicted values are obtained similarly, by replacing y and z in Equation 14.18 with py and pz, respectively. As in the previous section, four sets of starting values are required for the Adams-Bashforth-Moulton method. The first set comes directly from the initial conditions; the other three sets are obtained from the Runge-Kutta method.

Solved Problems

Solved Problem 14.1 Draw two solution curves to the differential equation $y' = 2y - x$.

A direction field for this equation is given by Figure 14-1. Two solution curves are also shown, one that passes through the point (0,0) and a second that passes through the point (0,2). Observe that each solution curve follows the flow of the line elements in the direction field.

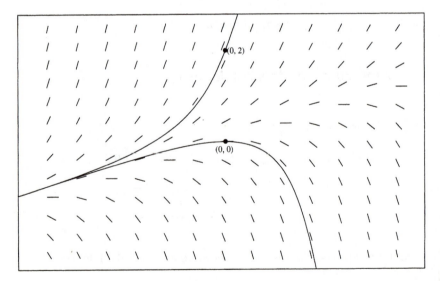

Figure 14-1

Solved Problem 14.2 Find $y(1)$ for $y' = y - x$; $y(0) = 2$, using Euler's method with $h = \frac{1}{4}$.

For this problem, $x_0 = 0$, $y_0 = 2$, and $f(x, y) = y - x$; so Equation 14.5 becomes $y'_n = y_n - x_n$. Because $h = \frac{1}{4}$,

$$x_1 = x_0 + h = \tfrac{1}{4} \quad x_2 = x_1 + h = \tfrac{1}{2} \quad x_3 = x_2 + h = \tfrac{3}{4} \quad x_4 = x_3 + h = 1$$

Using Equation 14.4 with $n = 0,1,2,3$ successively, we now compute the corresponding y-values.

$\mathbf{n = 0}$: $y_1 = y_0 + hy'_0$
But $y'_0 = f(x_0, y_0) = y_0 - x_0 = 2 - 0 = 2$
Hence, $y_1 = 2 + \frac{1}{4}(2) = \frac{5}{2}$

$\mathbf{n = 1}$: $y_2 = y_1 + hy'_1$
But $y'_1 = f(x_1, y_1) = y_1 - x_1 = \frac{5}{2} - \frac{1}{4} = \frac{9}{4}$
Hence, $y_2 = \frac{5}{2} + \frac{1}{4}(\frac{9}{4}) = \frac{49}{16}$

$\mathbf{n = 2}$: $y_3 = y_2 + hy'_2$
But $y'_2 = f(x_2, y_2) = y_2 - x_2 = \frac{49}{16} - \frac{1}{2} = \frac{41}{16}$
Hence, $y_3 = \frac{49}{16} + \frac{1}{4}(\frac{41}{16}) = \frac{237}{64}$

$\mathbf{n = 3}$: $y_4 = y_3 + hy'_3$
But $y'_3 = f(x_3, y_3) = y_3 - x_3 = \frac{237}{64} - \frac{3}{4} = \frac{189}{64}$
Hence, $y_4 = \frac{237}{64} + \frac{1}{4}(\frac{189}{64}) = \frac{1137}{256}$

Thus,

$$y(1) \approx y_4 = \tfrac{1137}{256} \approx 4.441$$

Note that the true solution is $Y(x) = e^x + x + 1$, so that $Y(1) \approx 4.718$. If we plot (x_n, y_n) for $n = 0,1,2,3$, and 4, and then connect successive points with straight line segments, as done in Figure 14-2, we have an approximation to the solution curve on $[0,1]$ for this initial-value problem.

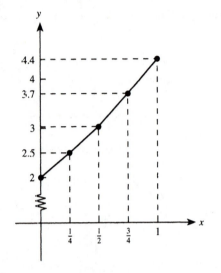

Figure 14-2

Chapter 15
BOUNDARY-VALUE PROBLEMS AND FOURIER SERIES

Second-Order Boundary-Value Problems

Standard Form

A boundary-value problem in standard form consists of the second-order linear differential equation

$$y'' + P(x)y' + Q(x)y = \phi(x) \qquad (15.1)$$

and the boundary conditions

$$\alpha_1 y(a) + \beta_1 y'(a) = \gamma_1$$
$$\alpha_2 y(b) + \beta_2 y'(b) = \gamma_2$$

(15.2)

where $P(x)$, $Q(x)$, and $\phi(x)$ are continuous in $[a, b]$ and α_1, α_2, β_1, β_2, γ_1, and γ_2 are all real constants. Furthermore, it is assumed that α_1 and β_1 are not both zero, and also that α_2 and β_2 are not both zero.

The boundary-value problem is said to be *homogeneous* if both the differential equation and the boundary conditions are homogeneous (i.e., $\phi(x) \equiv 0$ and $\gamma_1 = \gamma_2 = 0$). Otherwise the problem is *nonhomogeneous*. Thus a homogeneous boundary-value problem has the form

$$y'' + P(x)y' + Q(x)y = 0;$$
$$\alpha_1 y(a) + \beta_1 y'(a) = 0$$
$$\alpha_2 y(b) + \beta_2 y'(b) = 0$$

(15.3)

A somewhat more general homogeneous boundary-value problem than 15.3 is one where the coefficients $P(x)$ and $Q(x)$ also depend on an arbitrary constant λ. Such a problem has the form

$$y'' + P(x, \lambda)y' + Q(x, \lambda)y = 0;$$
$$\alpha_1 y(a) + \beta_1 y'(a) = 0$$
$$\alpha_2 y(b) + \beta_2 y'(b) = 0$$

(15.4)

Both 15.3 and 15.4 always admit the trivial solution $y(x) \equiv 0$.

Solutions

A boundary-value problem is solved by first obtaining the general solution to the differential equation, using any of the appropriate methods presented heretofore, and then applying the boundary conditions to evaluate the arbitrary constants.

Theorem 15.1. Let $y_1(x)$ and $y_2(x)$ be two linearly independent solutions of

$$y'' + P(x)y' + Q(x)y = 0$$

Nontrivial solutions (i.e., solutions not identically equal to zero) to the homogeneous boundary-value problem 15.3 exist if and only if the determinant

$$\begin{vmatrix} \alpha_1 y_1(a) + \beta_1 y_1'(a) & \alpha_1 y_2(a) + \beta_1 y_2'(a) \\ \alpha_2 y_1(b) + \beta_2 y_1'(b) & \alpha_2 y_2(b) + \beta_2 y_2'(b) \end{vmatrix} \tag{15.5}$$

equals zero.

Theorem 15.2. The nonhomogeneous boundary-value problem defined by 15.1 and 15.2 has a unique solution if and only if the associated homogeneous problem 15.3 has only the trivial solution.

In other words, *a nonhomogeneous problem has a unique solution when and only when the associated homogeneous problem has a unique solution.*

Eigenvalue Problems

When applied to the boundary-value problem 15.4, Theorem 15.1 shows that nontrivial solutions may exist for certain values of λ but not for other values of λ.

You Need to Know ✔

Those values of λ for which nontrivial solutions do exist are called *eigenvalues*; the corresponding nontrivial solutions are called *eigenfunctions*.

Sturm-Liouville Problems

A second-order *Sturm-Liouville* problem is a homogeneous boundary-value problem of the form

$$[p(x)y']' + q(x)y + \lambda w(x)y = 0; \tag{15.6}$$

$$\alpha_1 y(a) + \beta_1 y'(a) = 0$$
$$\alpha_2 y(b) + \beta_2 y'(b) = 0 \tag{15.7}$$

where $p(x)$, $p'(x)$, $q(x)$, and $w(x)$ are continuous on $[a, b]$, and both $p(x)$ and $w(x)$ are positive on $[a, b]$.

Equation 15.6 can be written in standard form 15.4 by dividing through by $p(x)$. Form 15.6, when attainable, is preferred, because Sturm-Liouville problems have desirable features not shared by more general eigenvalue problems. The second-order differential equation

$$a_2(x)y'' + a_1(x)y' + a_0(x)y + \lambda r(x)y = 0 \tag{15.8}$$

where $a_2(x)$ does not vanish on $[a, b]$, is equivalent to Equation 15.6 if and only if $a_2'(x) = a_1(x)$. This condition can always be forced by multiplying Equation 15.8 by a suitable factor.

Properties of Sturm-Liouville Problems

Property 15.1. The eigenvalues of a Sturm-Liouville problem are real and nonnegative.

Property 15.2. The eigenvalues of a Sturm-Liouville problem can be arranged to form a strictly increasing infinite sequence; that is, $0 \leq \lambda_1 < \lambda_2 < \lambda_3 < \cdots$. Furthermore, $\lambda_n \to \infty$ as $n \to \infty$.

Property 15.3. For each eigenvalue of a Sturm-Liouville problem, there exists one and only one linearly independent eigenfunction.

[By Property 15.3 there corresponds to each eigenvalue λ_n a unique eigenfunction where the coefficient on the lead term is one; we denote this eigenfunction by $e_n(x)$.]

Property 15.4. The set of eigenfunctions $\{e_1(x), e_2(x),...\}$ of a Sturm-Liouville problem satisfies the relation

$$\int_a^b w(x)e_n(x)e_m(x)dx = 0 \tag{15.9}$$

for $n \neq m$, where $w(x)$ is given in Equation 15.6.

Eigenfunction Expansions

A wide class of functions can be represented by infinite series of eigen-functions of a Sturm-Liouville problem.

Definition: A function $f(x)$ is *piecewise continuous on the open interval* $a < x < b$ if (1) $f(x)$ is continuous everywhere in $a < x < b$ with the possible exception of at most a *finite* number of points $x_1, x_2,..., x_n$, and (2) at these points of discontinuity, the right- and left-hand limits of $f(x)$ respectively $\lim_{\substack{x \to x_j \\ x > x_j}} f(x)$ and $\lim_{\substack{x \to x_j \\ x < x_j}} f(x)$, exist ($j = 1, 2,..., n$).

(Note that a continuous function is piecewise continuous.)

Definition: A function $f(x)$ is *piecewise continuous on the closed interval* $a \le x \le b$ if (1) it is piecewise continuous on the open interval $a < x < b$, (2) the right-hand limit of $f(x)$ exists at $x = a$ and (3) the left-hand limit of $f(x)$ exists at $x = b$.

Definition: A function $f(x)$ is *piecewise smooth* on $[a, b]$ if both $f(x)$ and $f'(x)$ are piecewise continuous on $[a,b]$.

Theorem 15.3. If $f(x)$ is piecewise smooth on $[a, b]$ and if $\{e_n(x)\}$ is the set of all eigenfunctions of a Sturm-Liouville problem (see Property 15.3), then

$$f(x) = \sum_{n=1}^{\infty} c_n e_n(x) \tag{15.10}$$

where

$$c_n = \frac{\int_a^b w(x) f(x) e_n(x) dx}{\int_a^b w(x) e_n^2(x) dx} \tag{15.11}$$

The representation 15.10 is valid at all points in the open interval (a, b) where $f(x)$ is continuous. The function $w(x)$ in 15.11 is given by Equation 15.6.

Because different Sturm-Liouville problems usually generate different sets of eigenfunctions, a given piecewise smooth function will have many expansions of the form 15.10. The basic features of all such expansions are exhibited by the trigonometric series discussed below.

Fourier Sine Series

The eigenfunctions of the Sturm-Liouville problem $y'' + \lambda y = 0$; $y(0) = 0$, $y(L) = 0$, where L is a real positive number, are $e_n(x) = \sin(n\pi x/L)$ ($n = 1,2,3,...$). Substituting these functions into 15.10, we obtain

$$f(x) = \sum_{n=1}^{\infty} c_n \sin \frac{n\pi x}{L} \tag{15.12}$$

For this Sturm-Liouville problem, $w(x) \equiv 1$, $a = 0$, and $b = L$; so that

$$\int_a^b w(x)e_n^2(x)dx = \int_0^L \sin^2 \frac{n\pi x}{L} dx = \frac{L}{2}$$

and 15.11 becomes

$$c_n = \frac{2}{L} \int_0^L f(x) \sin \frac{n\pi x}{L} dx \tag{15.13}$$

The expansion 15.12 with coefficients given by 15.13 is the *Fourier sine series* for $f(x)$ on $(0, L)$.

Fourier Cosine Series

The eigenfunctions of the Sturm-Liouville problem $y'' + \lambda y = 0$; $y'(0) = 0$, $y(L) = 0$, where L is a real positive number, are $e_0(x) = 1$ and $e_n(x) = \cos(n\pi x/L)$ ($n = 1,2,3,...$). Here $\lambda = 0$ is an eigenvalue with corresponding eigenfunction $e_0(x) = 1$. Substituting these functions into 15.10, where because of the additional eigenfunction $e_0(x)$ the summation now begins at $n = 0$, we obtain

$$f(x) = c_0 + \sum_{n=1}^{\infty} c_n \cos \frac{n\pi x}{L} \tag{15.14}$$

For this Sturm-Liouville problem, $w(x) \equiv 1$, $a = 0$, and $b = L$; so that

$$\int_a^b w(x)e_0^2(x)dx = \int_0^L dx = L$$

and

$$\int_a^b w(x)e_n^2(x)dx = \int_0^L \cos^2 \frac{n\pi x}{L} dx = \frac{L}{2}$$

Thus 15.11 becomes

$$c_0 = \frac{1}{L}\int_0^L f(x)dx \qquad c_n = \frac{2}{L}\int_0^L f(x)\cos\frac{n\pi x}{L}dx \qquad (15.15)$$

$$(n = 1, 2, \ldots)$$

The expansion 15.14 with coefficients given by 15.15 is the *Fourier cosine series* for $f(x)$ on $(0, L)$.

Solved Problems

Solved Problem 15.1 Solve $y'' + 2y' - 3y = 9x$; $y(0) = 1$, $y'(1) = 2$.

This is a nonhomogeneous boundary-value problem of forms 15.1 and 15.2, where $\phi(x) = 9x$, $\gamma_1 = 1$, and $\gamma_2 = 2$. The general solution to the homogeneous differential equation is $y = c_1 e^{-3x} + c_2 e^x$. If we apply homogeneous boundary conditions, we find that $c_1 = c_2 = 0$; hence the solution to the homogeneous problem is $y \equiv 0$. Since the associated homogeneous problem has only the trivial solution, it follows from Theorem 15.2 that the given problem has a unique solution. Solving the differential equation by the method of Chapter Six, we obtain

$$y = c_1 e^{-3x} + c_2 e^x - 3x - 2$$

Applying the boundary conditions, we find

$$c_1 + c_2 - 2 = 1 \quad -3c_1 e^{-3} + c_2 e - 3 = 2$$

whence

$$c_1 = \frac{3e - 5}{e + 3e^{-3}} \quad c_2 = \frac{5 + 9e^{-3}}{e + 3e^{-3}}$$

Finally,

$$y = \frac{(3e - 5)e^{-3x} + (5 + 9e^{-3})e^x}{e + 3e^{-3}} - 3x - 2$$

Solved Problem 15.2 Find the eigenvalues and eigenfunctions of

$$y'' - 4\lambda y' + 4\lambda^2 y = 0; \quad y(0) = 0, \quad y(1) + y'(1) = 0$$

The coefficients of the given differential equation are constants (with respect to x); hence, the general solution can be found by use of the characteristic equation. We write the characteristic equation in terms of the variable m, since λ now has another meaning. Thus we have $m^2 - 4\lambda m + 4\lambda^2 = 0$, which has the double root $m = 2\lambda$; the solution to the differential equation is $y = c_1 e^{2\lambda x} + c_2 x e^{2\lambda x}$. Applying the boundary conditions and simplifying, we obtain

$$c_1 = 0 \quad c_1(1 + 2\lambda) + c_2(2 + 2\lambda) = 0$$

It now follows that $c_1 = 0$ and either $c_2 = 0$ or $\lambda = -1$. The choice $c_2 = 0$ results in the trivial solution $y = 0$; the choice $\lambda = -1$ results in the nontrivial solution $y = c_2 x e^{2x}$, c_2 arbitrary. Thus, the boundary-value problem has the eigenvalue $\lambda = -1$ and the eigenfunction $y = c_2 x e^{2x}$.

Solved Problem 15.3 Find a Fourier sine series for $f(x) = \begin{cases} 0 & x \leq 2 \\ 2 & x > 2 \end{cases}$ on $(0,3)$.

Using Equation 15.13 with $L = 3$, we obtain

$$c_n = \frac{2}{3} \int_0^3 f(x) \sin \frac{n\pi x}{3} dx$$

$$= \frac{2}{3} \int_0^2 (0) \sin \frac{n\pi x}{3} dx + \frac{2}{3} \int_2^3 (2) \sin \frac{n\pi x}{3} dx$$

$$= 0 + \frac{4}{3} \left[-\frac{3}{n\pi} \cos \frac{n\pi x}{3} \right]_{x=2}^{x=3} = \frac{4}{n\pi} \left[\cos \frac{2n\pi}{3} - \cos n\pi \right]$$

Thus Equation 15.12 becomes

$$f(x) = \sum_{n=1}^{\infty} \frac{4}{n\pi} \left[\cos\frac{2n\pi}{3} - (-1)^n \right] \sin\frac{n\pi x}{3}$$

Furthermore, $\cos\dfrac{2\pi}{3} = -\dfrac{1}{2},\ \cos\dfrac{4\pi}{3} = -\dfrac{1}{2},\ \cos\dfrac{6\pi}{3} = 1,\ldots$

Hence,

$$f(x) = \frac{4}{\pi} \left(\frac{1}{2}\sin\frac{\pi x}{3} - \frac{3}{4}\sin\frac{2\pi x}{3} + \frac{2}{3}\sin\frac{3\pi x}{3} - \cdots \right)$$

Since $f(x)$ is piecewise smooth on $[0,3]$ and continuous everywhere in $(0,3)$ except at $x = 2$, it follows from Theorem 15.3 that this equality is valid everywhere in $(0,3)$ except at $x = 2$.

Appendix
LAPLACE
TRANSFORMS

	$f(x)$	$F(s) = \mathcal{L}\{f(x)\}$
1.	1	$\dfrac{1}{s} \quad (s > 0)$
2.	x	$\dfrac{1}{s^2} \quad (s > 0)$
3.	$x^{n-1} \quad (n = 1,2,\ldots)$	$\dfrac{(n-1)!}{s^n} \quad (s > 0)$
4.	$\sqrt{x}$	$\dfrac{1}{2}\sqrt{\pi}\, s^{-3/2} \quad (s > 0)$
5.	$1/\sqrt{x}$	$\sqrt{\pi}\, s^{-1/2} \quad (s > 0)$
6.	$x^{n-1/2} \quad (n = 1,2,\ldots)$	$\dfrac{(1)(3)(5)\cdots(2n-1)\sqrt{\pi}}{2^n} s^{-n-1/2}$ $(s > 0)$
7.	e^{ax}	$\dfrac{1}{s-a} \quad (s > a)$
8.	$\sin ax$	$\dfrac{a}{s^2 + a^2} \quad (s > 0)$
9.	$\cos ax$	$\dfrac{s}{s^2 + a^2} \quad (s > 0)$
10.	$\sinh ax$	$\dfrac{a}{s^2 - a^2} \quad (s > |a|)$

	$f(x)$	$F(s) = \mathcal{L}\{f(x)\}$
11.	$\cosh ax$	$\dfrac{s}{s^2 - a^2} \quad (s > |a|)$
12	$x \sin ax$	$\dfrac{2as}{(s^2 + a^2)^2} \quad (s > 0)$
13.	$x \cos ax$	$\dfrac{s^2 - a^2}{(s^2 + a^2)^2} \quad (s > 0)$
14.	$x^{n-1} e^{ax} \ (n = 1, 2, \ \ldots)$	$\dfrac{(n-1)!}{(s-a)^n} \quad (s > a)$
15.	$e^{bx} \sin ax$	$\dfrac{a}{(s-b)^2 + a^2} \quad (s > b)$
16.	$e^{bx} \cos ax$	$\dfrac{s - b}{(s-b)^2 + a^2} \quad (s > b)$
17.	$\sin ax - ax \cos ax$	$\dfrac{2a^3}{(s^2 + a^2)^2} \quad (s > 0)$
18.	$\dfrac{1}{a} e^{-x/a}$	$\dfrac{1}{1 + as}$
19.	$\dfrac{1}{a}(e^{ax} - 1)$	$\dfrac{1}{s(s - a)}$
20.	$1 - e^{-x/a}$	$\dfrac{1}{s(1 + as)}$
21.	$\dfrac{1}{a^2} x e^{-x/a}$	$\dfrac{1}{(1 + as)^2}$

	$f(x)$	$F(s) = \mathcal{L}\{f(x)\}$
22.	$\dfrac{e^{ax} - e^{bx}}{a - b}$	$\dfrac{1}{(s-a)(s-b)}$
23.	$\dfrac{e^{-x/a} - e^{-x/b}}{a - b}$	$\dfrac{1}{(1+as)(1+bs)}$
24.	$(1 + ax)e^{ax}$	$\dfrac{s}{(s-a)^2}$
25.	$\dfrac{1}{a^3}(a - x)e^{-x/a}$	$\dfrac{s}{(1+as)^2}$
26.	$\dfrac{ae^{ax} - be^{bx}}{a - b}$	$\dfrac{s}{(s-a)(s-b)}$
27.	$\dfrac{ae^{-x/b} - be^{-x/a}}{ab(a - b)}$	$\dfrac{s}{(1+as)(1+bs)}$
28.	$\dfrac{1}{a^2}(e^{ax} - 1 - ax)$	$\dfrac{1}{s^2(s-a)}$
29.	$\sin^2 ax$	$\dfrac{2a^2}{s(s^2 + 4a^2)}$
30.	$\sinh^2 ax$	$\dfrac{2a^2}{s(s^2 - 4a^2)}$
31.	$\dfrac{1}{\sqrt{2}}\left(\cosh\dfrac{ax}{\sqrt{2}}\sin\dfrac{ax}{\sqrt{2}} - \sinh\dfrac{ax}{\sqrt{2}}\cos\dfrac{ax}{\sqrt{2}}\right)$	$\dfrac{a^3}{s^4 + a^4}$
32.	$\sin\dfrac{ax}{\sqrt{2}}\sinh\dfrac{ax}{\sqrt{2}}$	$\dfrac{a^2 s}{s^4 + a^4}$

	$f(x)$	$F(s) = \mathcal{L}\{f(x)\}$
33.	$\dfrac{1}{\sqrt{2}}\left(\begin{array}{l}\cos\dfrac{ax}{\sqrt{2}}\sinh\dfrac{ax}{\sqrt{2}}\ + \\ \sin\dfrac{ax}{\sqrt{2}}\cosh\dfrac{ax}{\sqrt{2}}\end{array}\right)$	$\dfrac{as^2}{s^4 + a^4}$
34.	$\cos\dfrac{ax}{\sqrt{2}}\cosh\dfrac{ax}{\sqrt{2}}$	$\dfrac{s^3}{s^4 + a^4}$
35.	$\dfrac{1}{2}(\sinh ax - \sin ax)$	$\dfrac{a^3}{s^4 - a^4}$
36.	$\dfrac{1}{2}(\cosh ax - \cos ax)$	$\dfrac{a^2 s}{s^4 - a^4}$
37.	$\dfrac{1}{2}(\sinh ax + \sin ax)$	$\dfrac{as^2}{s^4 - a^4}$
38.	$\dfrac{1}{2}(\cosh ax + \cos ax)$	$\dfrac{s^3}{s^4 - a^4}$
39.	$\sin ax \sinh ax$	$\dfrac{2a^2 s}{s^4 + 4a^4}$
40.	$\cos ax \sinh ax$	$\dfrac{a(s^2 - 2a^2)}{s^4 + 4a^4}$
41.	$\sin ax \cosh ax$	$\dfrac{a(s^2 + 2a^2)}{s^4 + 4a^4}$
42.	$\cos ax \cosh ax$	$\dfrac{s^3}{s^4 + 4a^4}$
43.	$\dfrac{1}{2}(\sin ax + ax \cos ax)$	$\dfrac{as^2}{(s^2 + a^2)^2}$

	$f(x)$	$F(s) = \mathcal{L}\{f(x)\}$
44.	$\cos ax - \dfrac{ax}{2}\sin ax$	$\dfrac{s^3}{(s^2 + a^2)^2}$
45.	$\dfrac{1}{2}(ax\cosh ax - \sinh ax)$	$\dfrac{a^3}{(s^2 - a^2)^2}$
46.	$\dfrac{x}{2}\sinh ax$	$\dfrac{as}{(s^2 - a^2)^2}$
47.	$\dfrac{1}{2}(\sinh ax + ax\cosh ax)$	$\dfrac{as^2}{(s^2 - a^2)^2}$
48.	$\cosh ax + \dfrac{ax}{2}\sinh ax$	$\dfrac{s^3}{(s^2 - a^2)^2}$
49.	$\dfrac{a\sin bx - b\sin ax}{a^2 - b^2}$	$\dfrac{ab}{(s^2 + a^2)(s^2 + b^2)}$
50.	$\dfrac{\cos bx - \cos ax}{a^2 - b^2}$	$\dfrac{s}{(s^2 + a^2)(s^2 + b^2)}$
51.	$\dfrac{a\sin ax - b\sin bx}{a^2 - b^2}$	$\dfrac{s^2}{(s^2 + a^2)(s^2 + b^2)}$
52.	$\dfrac{a^2\cos ax - b^2\cos bx}{a^2 - b^2}$	$\dfrac{s^3}{(s^2 + a^2)(s^2 + b^2)}$
53.	$\dfrac{b\sinh ax - a\sinh bx}{a^2 - b^2}$	$\dfrac{ab}{(s^2 - a^2)(s^2 - b^2)}$
54.	$\dfrac{\cosh ax - \cosh bx}{a^2 - b^2}$	$\dfrac{s}{(s^2 - a^2)(s^2 - b^2)}$

	$f(x)$	$F(s) = \mathcal{L}\{f(x)\}$
55.	$\dfrac{a \sinh ax - b \sinh bx}{a^2 - b^2}$	$\dfrac{s^2}{(s^2 - a^2)(s^2 - b^2)}$
56.	$\dfrac{a^2 \cosh ax - b^2 \cosh bx}{a^2 - b^2}$	$\dfrac{s^3}{(s^2 - a^2)(s^2 - b^2)}$
57.	$x - \dfrac{1}{a} \sin ax$	$\dfrac{a^2}{s^2(s^2 + a^2)}$
58.	$\dfrac{1}{a} \sinh ax - x$	$\dfrac{a^2}{s^2(s^2 - a^2)}$
59.	$1 - \cos ax - \dfrac{ax}{2} \sin ax$	$\dfrac{a^4}{s(s^2 + a^2)^2}$
60.	$1 - \cosh ax + \dfrac{ax}{2} \sinh ax$	$\dfrac{a^4}{s(s^2 - a^2)^2}$
61.	$1 + \dfrac{b^2 \cos ax - a^2 \cos bx}{a^2 - b^2}$	$\dfrac{a^2 b^2}{s(s^2 + a^2)(s^2 + b^2)}$
62.	$1 + \dfrac{b^2 \cosh ax - a^2 \cosh bx}{a^2 - b^2}$	$\dfrac{a^2 b^2}{s(s^2 - a^2)(s^2 - b^2)}$
63.	$\dfrac{1}{8}\left[(3 - a^2 x^2) \sin ax - 3ax \cos ax \right]$	$\dfrac{a^5}{(s^2 + a^2)^3}$
64.	$\dfrac{x}{8}\left[\sin ax - ax \cos ax \right]$	$\dfrac{a^3 s}{(s^2 + a^2)^3}$
65.	$\dfrac{1}{8}\left[(1 + a^2 x^2) \sin ax - ax \cos ax \right]$	$\dfrac{a^3 s^2}{(s^2 + a^2)^3}$

	$f(x)$	$F(s) = \mathcal{L}\{f(x)\}$
66.	$\dfrac{1}{8}\left[(3+a^2x^2)\sinh ax\ -\ 3ax\cosh ax\right]$	$\dfrac{a^5}{(s^2-a^2)^3}$
67.	$\dfrac{x}{8}(ax\cosh ax - \sinh ax)$	$\dfrac{a^3 s}{(s^2-a^2)^3}$
68.	$\dfrac{1}{8}\left[\begin{array}{l}ax\cosh ax\ -\\ (1-a^2x^2)\sinh ax\end{array}\right]$	$\dfrac{a^3 s^2}{(s^2-a^2)^3}$
69.	$\dfrac{1}{n!}(1-e^{-x/a})^n$	$\dfrac{1}{s(as+1)(as+2)\cdots(as+n)}$
70.	$\sin(ax+b)$	$\dfrac{s\sin b + a\cos b}{s^2+a^2}$
71.	$\cos(ax+b)$	$\dfrac{s\cos b - a\sin b}{s^2+a^2}$
72.	$e^{-ax} - e^{ax/2}\left(\cos\dfrac{ax\sqrt{3}}{2}\ -\ \sqrt{3}\sin\dfrac{ax\sqrt{3}}{2}\right)$	$\dfrac{3a^2}{s^3+a^3}$
73.	$\dfrac{1+2ax}{\sqrt{\pi x}}$	$\dfrac{s+a}{s\sqrt{s}}$
74.	$e^{-ax}/\sqrt{\pi x}$	$\dfrac{1}{\sqrt{s+a}}$
75.	$\dfrac{1}{2x\sqrt{\pi x}}(e^{bx}-e^{ax})$	$\sqrt{s-a}-\sqrt{s-b}$
76.	$\dfrac{1}{\sqrt{\pi x}}\cos 2\sqrt{ax}$	$\dfrac{1}{\sqrt{s}}e^{-a/s}$

	$f(x)$	$F(s) = \mathcal{L}\{f(x)\}$
77.	$\dfrac{1}{\sqrt{\pi x}}\cosh 2\sqrt{ax}$	$\dfrac{1}{\sqrt{s}}e^{a/s}$
78.	$\dfrac{1}{\sqrt{a\pi}}\sin 2\sqrt{ax}$	$s^{-3/2}e^{-a/s}$
79.	$\dfrac{1}{\sqrt{a\pi}}\sinh 2\sqrt{ax}$	$s^{-3/2}e^{a/s}$
80.	$J_0(2\sqrt{ax})$	$\dfrac{1}{s}e^{-a/s}$
81.	$\sqrt{x/a}\,J_1(2\sqrt{ax})$	$\dfrac{1}{s^2}e^{-a/s}$
82.	$(x/a)^{(p-1)/2}J_{p-1}(2\sqrt{ax})$ $(p > 0)$	$s^{-p}e^{-a/s}$
83.	$J_0(x)$	$\dfrac{1}{\sqrt{s^2+1}}$
84.	$J_1(x)$	$\dfrac{\sqrt{s^2+1}-s}{\sqrt{s^2+1}}$
85.	$J_p(x)\quad(p > -1)$	$\dfrac{(\sqrt{s^2+1}-s)^p}{\sqrt{s^2+1}}$
86.	$x^p J_p(ax)\quad\left(p > -\dfrac{1}{2}\right)$	$\dfrac{(2a)^p\,\Gamma(p+\frac{1}{2})}{\sqrt{\pi}\,(s^2+a^2)^{p+(1/2)}}$
87.	$\dfrac{x^{p-1}}{\Gamma(p)}\quad(p > 0)$	$\dfrac{1}{s^p}$
88.	$\dfrac{4^n n!}{(2n)!\sqrt{\pi}}x^{n-(1/2)}$	$\dfrac{1}{s^n\sqrt{s}}$

	$f(x)$	$F(s) = \mathcal{L}\{f(x)\}$
89.	$\dfrac{x^{p-1}}{\Gamma(p)}e^{-ax} \qquad (p > 0)$	$\dfrac{1}{(s+a)^p}$
90.	$\dfrac{1 - e^{ax}}{x}$	$\ln\dfrac{s-a}{s}$
91.	$\dfrac{e^{bx} - e^{ax}}{x}$	$\ln\dfrac{s-a}{s-b}$
92.	$\dfrac{2}{x}\sinh ax$	$\ln\dfrac{s+a}{s-a}$
93.	$\dfrac{2}{x}(1 - \cos ax)$	$\ln\dfrac{s^2 + a^2}{s^2}$
94.	$\dfrac{2}{x}(\cos bx - \cos ax)$	$\ln\dfrac{s^2 + a^2}{s^2 + b^2}$
95.	$\dfrac{\sin ax}{x}$	$\arctan\dfrac{a}{s}$
96.	$\dfrac{2}{x}\sin ax \cos bx$	$\arctan\dfrac{2as}{s^2 - a^2 + b^2}$
97.	$\sin |ax|$	$\left(\dfrac{a}{s^2 + a^2}\right)\left(\dfrac{1 + e^{-(\pi/a)s}}{1 - e^{-(\pi/a)s}}\right)$

Index

133